AN INTRODUCTION TO BOOTSTRAP METHODS WITH APPLICATIONS TO R

AN INTRODUCTION TO BOOTSTRAP METHODS WITH APPLICATIONS TO R

Michael R. Chernick
Lankenau Institute for Medical Research, Wynnewood, PA
Thomas Jefferson University, Philadelphia, PA

Robert A. LaBudde
Least Cost Formulations Ltd., Norfolk, VA
Old Dominion University, Norfolk, VA

A JOHN WILEY & SONS, INC., PUBLICATION

For general information on our other products and services or for technical support, please contact our Customer Care Department within the United States at (800) 762-2974, outside the United States at (317) 572-3993 or fax (317) 572-4002.

Wiley also publishes its books in a variety of electronic formats. Some content that appears in print may not be available in electronic formats. For more information about Wiley products, visit our web site at www.wiley.com.

Library of Congress Cataloging-in-Publication Data:

Chernick, Michael R.
    An introduction to bootstrap methods with applications to R / Michael R. Chernick, Robert A. LaBudde.
        p. cm.
    Includes bibliographical references and index.
    ISBN 978-0-470-46704-6 (hardback)
1. Bootstrap (Statistics)   2. R (Computer program language)   I. LaBudde, Robert A., 1947–   II. Title.
    QA276.8.C478 2011
    519.5'4–dc22

                                                                                2011010972

Printed in the United States of America.

10 9 8 7 6 5 4 3 2 1

CONTENTS

PREFACE

The term "bootstrapping" refers to the concept of "pulling oneself up by one's bootstraps," a phrase apparently first used in *The Singular Travels, Campaigns and Adventures of Baron Munchausen* by Rudolph Erich Raspe in 1786. The derivative of the same term is used in a similar manner to describe the process of "booting" a computer by a sequence of software increments loaded into memory at power-up.

In statistics, "bootstrapping" refers to making inferences about a sampling distribution of a statistic by "resampling" the sample itself with replacement, as if it were a finite population. To the degree that the resampling distribution mimics the original sampling distribution, the inferences are accurate. The accuracy improves as the size of the original sample increases, if the central limit theorem applies.

"Resampling" as a concept was first used by R. A. Fisher (1935) in his famous randomization test, and by E. J. G. Pitman (1937, 1938), although in these cases the sampling was done without replacement.

The "bootstrap" as sampling with replacement and its Monte Carlo approximate form was first presented in a Stanford University technical report by Brad Efron in 1977. This report led to his famous paper in the *Annals of Statistics* in 1979. However, the Monte Carlo approximation may be much older. In fact, it is known that Julian Simon at the University of Maryland proposed the Monte Carlo approximation as an educational tool for teaching probability and statistics. In the 1980s, Simon and Bruce started a company called Resampling Stats that produced a software product to do bootstrap and permutation sampling for both educational and inference purposes.

But it was not until Efron's paper that related the bootstrap to the jackknife and other resampling plans that the statistical community got involved. Over the next 20 years, the theory and applications of the bootstrap blossomed, and the Monte Carlo approximation to the bootstrap became a very practiced approach to making statistical inference without strong parametric assumptions.

Michael Chernick was a graduate student in statistics at the time of Efron's early research and saw the development of bootstrap methods from its very beginning. However, Chernick did not get seriously involved into bootstrap research until 1984 when he started to find practical applications in nonlinear regression models and classification problems while employed at the Aerospace Corporation.

After meeting Philip Good in the mid-1980s, Chernick and Good set out to accumulate an extensive bibliography on resampling methods and planned a two-volume text with Chernick to write the bootstrap methods volume and Good the volume on permutation tests. The project that was contracted by Oxford University Press was eventually abandoned. Good eventually published his work with Springer-Verlag, and later, Chernick separately published his work with Wiley. Chernick and Good later taught together short courses on resampling methods, first at UC Irvine and later the Joint Statistical Meetings in Indianapolis in 2000. Since that time, Chernick has taught resampling methods with Peter Bruce and later bootstrap methods for statistics.com.

Robert LaBudde wrote his PhD dissertation in theoretical chemistry on the application of Monte Carlo methods to simulating elementary chemical reactions. A long time later, he took courses in resampling methods and bootstrap methods from statistics.com, and it was in this setting in Chernick's bootstrap methods course that the two met. In later sessions, LaBudde was Chernick's teaching assistant and collaborator in the course and provided exercises in the R programming language. This course was taught using the second edition of Chernick's bootstrap text. However, there were several deficiencies with the use of this text including lack of homework problems and software to do the applications. The level of knowledge required was also variable. This text is intended as a text for an elementary course in bootstrap methods including Chernick's statistics.com course.

This book is organized in a similar way as *Bootstrap Methods: A Guide for Practitioners and Researchers*. Chapter 1 provides an introduction with some historical background, a formal description of the bootstrap and its relationship to other resampling methods, and an overview of the wide variety of applications of the approach. An introduction to R programming is also included to prepare the student for the exercises and applications that require programming using this software system. Chapter 2 covers point estimation, Chapter 3 confidence intervals, and Chapter 4 hypothesis testing. More advanced topics begin with time series in Chapter 5. Chapter 6 covers some of the more important variants of the bootstrap. Chapter 7 covers special topics including spatial data analysis, *P*-value adjustment in multiple testing, censored data, subset selection in regression models, process capability indices, and some new material on bioequivalence and covariate adjustment to area under the curve for receiver operating characteristics for diagnostic tests. The final chapter, Chapter 8, covers various examples where the bootstrap was found not to work as expected (fails asymptotic consistency requirements). But in every case, modifications have been found that are consistent.

This text is suitable for a one-semester or one-quarter introductory course in bootstrap methods. It is designed for users of statistics more than statisticians. So, students with an interest in engineering, biology, genetics, geology, physics, and even psychology and other social sciences may be interested in this course because of the various applications in their field. Of course, statisticians needing a basic understanding of the bootstrap and the surrounding literature may find the course useful. But it is not intended for a graduate course in statistics such as Hall (1992), Shao and Tu (1995), or Davison and Hinkley (1997). A shorter introductory course could be taught using just Chapters 1–4. Chapters 1–4 could also be used to incorporate bootstrap methods into a first course on statistical inference. References to the literature are covered in the historical notes sections in each chapter. At the end of each chapter is a set of homework exercises that the instructor may select from for homework assignments.

Initially, it was our goal to create a text similar to Chernick (2007) but more suitable for a full course on bootstrapping with a large number of exercises and examples illustrated in R. Also, the intent was to make the course elementary with technical details left for the interested student to read the original articles or other books. Our belief was that there were few new developments to go beyond the coverage of Chernick (2007). However, we found that with the introduction of "bagging" and "boosting," a new role developed for the bootstrap, particularly when estimating error rates for classification problems. As a result, we felt that it was appropriate to cover the new topics and applications in more detail. So parts of this text are not at the elementary level.

MICHAEL R. CHERNICK
ROBERT A. LABUDDE

An accompanying website has been established for this book. Please visit: http://lcfltd.com/Downloads/IntroBootstrap/IntroBootstrap.zip.

ACKNOWLEDGMENTS

The authors would like to thank our acquisitions editor Steve Quigley, who has always been enthusiastic about our proposals and always provides good advice. Also, Jackie Palmieri at Wiley for always politely reminding us when our manuscript was expected and for cheerfully accepting changes when delays become necessary. But it is that gentle push that gets us moving to completion. We also thank Dr. Jiajing Sun for her review and editing of the manuscript as well as her help putting the equations into Latex. We especially would like to thank Professor Dmitiris Politis for a nice and timely review of the entire manuscript. He also provided us with several suggestions for improving the text and some additions to the literature to take account of important ideas that we omitted. He also provided us with a number of reference that preceded Bickel, Gotze, and van Zwet (1996) on the value of taking bootstrap samples of size m less than n, as well as some other historical details for sieve bootstrap, and subsampling.

M.R.C.
R.A.L.

LIST OF TABLES

1

INTRODUCTION

1.1 HISTORICAL BACKGROUND

The "bootstrap" is one of a number of techniques that is now part of the broad umbrella of nonparametric statistics that are commonly called resampling methods. Some of the techniques are far older than the bootstrap. Permutation methods go back to Fisher (1935) and Pitman (1937, 1938), and the jackknife started with Quenouille (1949). Bootstrapping was made practical through the use of the Monte Carlo approximation, but it too goes back to the beginning of computers in the early 1940s.

However, 1979 is a critical year for the bootstrap because that is when Brad Efron's paper in the *Annals of Statistics* was published (Efron, 1979). Efron had defined a resampling procedure that he coined as bootstrap. He constructed it as a simple approximation to the jackknife (an earlier resampling method that was developed by John Tukey), and his original motivation was to derive properties of the bootstrap to better understand the jackknife. However, in many situations, the bootstrap is as good as or better than the jackknife as a resampling procedure. The jackknife is primarily useful for small samples, becoming computationally inefficient for larger samples but has become more feasible as computer speed increases. A clear description of the jackknife

An Introduction to Bootstrap Methods with Applications to R, First Edition. Michael R. Chernick, Robert A. LaBudde.
© 2011 John Wiley & Sons, Inc. Published 2011 by John Wiley & Sons, Inc.

and its connecton to the bootstrap can be found in the SIAM monograph Efron (1982). A description of the jackknife is also given in Section 1.2.1.

Although permutation tests were known in the 1930s, an impediment to their use was the large number (i.e., $n!$) of distinct permutations available for samples of size n. Since ordinary bootstrapping involves sampling with replacement n times for a sample of size n, there are n^n possible distinct ordered bootstrap samples (though some are equivalent under the exchangeability assumption because they are permutations of each other). So, complete enumeration of all the bootstrap samples becomes infeasible except in very small sample sizes. Random sampling from the set of possible bootstrap samples becomes a viable way to approximate the distribution of bootstrap samples. The same problem exists for permutations and the same remedy is possible. The only difference is that $n!$ does not grow as fast as n^n, and complete enumeration of permutations is possible for larger n than for the bootstrap.

The idea of taking several Monte Carlo samples of size n with replacement from the original observations was certainly an important idea expressed by Efron but was clearly known and practiced prior to Efron (1979). Although it may not be the first time it was used, Julian Simon laid claim to priority for the bootstrap based on his use of the Monte Carlo approximation in Simon (1969). But Simon was only recommending the Monte Carlo approach as a way to teach probability and statistics in a more intuitive way that does not require the abstraction of a parametric probability model for the generation of the original sample. After Efron made the bootstrap popular, Simon and Bruce joined the campaign (see Simon and Bruce, 1991, 1995).

Efron, however, starting with Efron (1979), first connected bootstrapping to the jackknife, delta method, cross-validation, and permutation tests. He was the first to show it to be a real competitor to the jackknife and delta method for estimating the standard error of an estimator. Also, quite early on, Efron recognized the broad applicability of bootstrapping for confidence intervals, hypothesis testing, and more complex problems. These ideas were emphasized in Efron and Gong (1983), Diaconis and Efron (1983), Efron and Tibshirani (1986), and the SIAM monograph (Efron 1982). These influential articles along with the SIAM monograph led to a great deal of research during the 1980s and 1990s. The explosion of bootstrap papers grew at an exponential rate. Key probabilistic results appeared in Singh (1981), Bickel and Freedman (1981, 1984), Beran (1982), Martin (1990), Hall (1986, 1988), Hall and Martin (1988), and Navidi (1989).

In a very remarkable paper, Efron (1983) used simulation comparisons to show that the use of bootstrap bias correction could provide better estimates of classification error rate than the very popular cross-validation approach (often called leave-one-out and originally proposed by Lachenbruch and Mickey, 1968). These results applied when the sample size was small, and classification was restricted to two or three classes only, and the predicting features had multivariate Gaussian distributions. Efron compared several variants of the bootstrap with cross-validation and the resubstitution methods. This led to several follow-up articles that widened the applicability and superiority of a version of the bootstrap called 632. See Chatterjee and Chatterjee (1983), Chernick et al. (1985, 1986, 1988a, b), Jain et al. (1987), and Efron and Tibshirani (1997).

Chernick was a graduate student at Stanford in the late 1970s when the bootstrap activity began on the Stanford and Berkeley campuses. However, oddly the bootstrap did not catch on with many graduate students. Even Brad Efron's graduate students chose other topics for their dissertation. Gail Gong was the first student of Efron to do a dissertation on the bootstrap. She did very useful applied work on using the bootstrap in model building (particularly for logistic regression subset selection). See Gong (1986). After Gail Gong, a number of graduate students wrote dissertations on the bootstrap under Efron, including Terry Therneau, Rob Tibshirani, and Tim Hesterberg. Michael Martin visited Stanford while working on his dissertation on bootstrap confidence intervals under Peter Hall. At Berkeley, William Navidi did his thesis on bootstrapping in regression and econometric models under David Freedman.

While exciting theoretical results developed for the bootstrap in the 1980s and 1990s, there were also negative results where it was shown that the bootstrap estimate is not "consistent" in the probabilistic sense (i.e., approaches the true parameter value as the sample size becomes infinite). Examples included the mean when the population distribution does not have a finite variance and when the maximum or minimum is taken from a sample. This is illustrated in Athreya (1987a, b), Knight (1989), Angus (1993), and Hall et al. (1993). The first published example of an inconsistent bootstrap estimate appeared in Bickel and Freedman (1981). Shao et al. (2000) showed that a particular approach to bootstrap estimation of individual bioequivalence is also inconsistent. They also provide a modification that is consistent. Generally, the bootstrap is consistent when the central limit theorem applies (a sufficient condition is Lyapanov's condition that requires existence of the $2 + \delta$ moment of the population distribution). Consistency results in the literature are based on the existence of Edgeworth expansions; so, additional smoothness conditions for the expansion to exist have also been assumed (but it is not known whether or not they are necessary).

One extension of the bootstrap called m-out-of-n was suggested by Bickel and Ren (1996) in light of previous research on it, and it has been shown to be a method to overcome inconsistency of the bootstrap in several instances. In the m-out-of-n bootstrap, sampling is with replacement from the original sample but with a value of m that is smaller than n. See Bickel et al. (1997), Gine and Zinn (1989), Arcones and Gine (1989), Fukuchi (1994), and Politis et al. (1999).

Some bootstrap approaches in time series have been shown to be inconsistent. Lahiri (2003) covered the use of bootstrap in time series and other dependent cases. He showed that there are remedies for the m-dependent and moving block bootstrap cases (see Section 5.5 for some coverage of moving block bootstrap) that are consistent.

1.2 DEFINITION AND RELATIONSHIP TO THE DELTA METHOD AND OTHER RESAMPLING METHODS

We will first provide an informal definition of bootstrap to provide intuition and understanding before a more formal mathematical definition. The objective of bootstrapping is to estimate a parameter based on the data, such as a mean, median, or standard deviation. We are also interested in the properties of the distribution for the parameter's estimate and may want to construct confidence intervals. But we do not want to make overly restrictive assumptions about the form of the distribution that the observed data came from.

For the simple case of independent observations coming from the same population distribution, the basic element for bootstrapping is the empirical distribution. The empirical distribution is just the discrete distribution that gives equal weight to each data point (i.e., it assigns probability $1/n$ to each of the original n observations and shall be denoted F_n).

Most of the common parameters that we consider are functionals of the unknown population distribution. A functional is simply a mapping that takes a function F into a real number. In our case, we are only interested in the functionals of cumulative probability distribution functions. So, for example, the mean and variance of a distribution can be represented as functionals in the following way. Let μ be the mean for a distribution function F, then $\mu = \int x dF(x)$ Let σ^2 be the variance then $\sigma^2 = \int (x - \mu)^2 dF(x)$. These integrals over the entire possible set of x values in the domain of F are particular examples of functionals. It is interesting that the sample estimates most commonly used for these parameters are the same functionals applied to the F_n.

Now the idea of bootstrap is to use only what you know from the data and not introduce extraneous assumptions about the population distribution. The "bootstrap principle" says that when F is the population distribution and $T(F)$ is the functional that defines the parameter, we wish to estimate based on a sample of size n, let F_n play the role of F and F_n^*, the bootstrap distribution (soon to be defined), play the role of F_n in the resampling process. Note that the original sample is a sample of n independent identically distributed observations from the distribution F and the sample estimate of the parameter is $T(F_n)$. So, in bootstrapping, we let F_n play the role of F and take n independent and identically distributed observations from F_n. Since F_n is the empirical distribution, this is just sampling randomly with replacement from the original data.

Suppose we have $n = 5$ and the observations are $X_1 = 7$, $X_2 = 5$, $X_3 = 3$, $X_4 = 9$, and $X_5 = 6$ and that we are estimating the mean. Then, the sample estimate of the population parameter is the sample mean, $(7 + 5 + 3 + 9 + 6)/5 = 6.0$. Then sampling from the data with replacement generates what we call a bootstrap sample.

The bootstrap sample is denoted X_1^*, X_2^*, X_3^*, X_4^*, and X_5^*. The distribution for sampling with replacement from F_n is called the bootstrap distribution, which we previously denoted by F_n^*. The bootstrap estimate is then $T(F_n^*)$. So a bootstrap sample might be $X_1^* = 5$, $X_2^* = 9$, $X_3^* = 7$, $X_4^* = 7$, and $X_5^* = 5$, with estimate $(5 + 9 + 7 + 7 + 5)/5 = 6.6$.

Note that, although it is possible to get the original sample back typically some values get repeated one or more times and consequently others get omitted. For this bootstrap sample, the bootstrap estimate of the mean is $(5 + 9 + 7 + 7 + 5)/5 = 6.6$. Note that the bootstrap estimate differs from the original sample estimate, 6.0. If we take another bootstrap sample, we may get yet another estimate that may be different from the previous one and the original sample. Assume for the second bootstrap sample we get in this case the observation equal to 9 repeated once. Then, for this bootstrap sample, $X_1^* = 9$, $X_2^* = 9$, $X_3^* = 6$, $X_4^* = 7$, and $X_5^* = 5$, and the bootstrap estimate for the mean is 7.2.

If we repeat this many times, we get a histogram of values for the mean, which we will call the Monte Carlo approximation to the bootstrap distribution. The average of all these values will be very close to 6.0 since the theoretical mean of the bootstrap distribution is the sample mean. But from the histogram (i.e., resampling distribution), we can also see the variability of these estimates and can use the histogram to estimate skewness, kurtosis, standard deviation, and confidence intervals.

In theory, the exact bootstrap estimate of the parameter could be calculated by averaging appropriately over all possible bootstrap samples, and in this example for the mean, that value would be 6.0. As noted before, there can be n^n distinct bootstrap samples (taking account of the ordering of the observations), and so even for $n = 10$, this becomes very large (i.e., 10 billion). So, in practice, a Monte Carlo approximation is used.

If you randomly generate $M = 10,000$ or $100,000$ bootstrap samples, the distribution of bootstrap estimates will approximate the bootstrap distribution for the estimate. The larger M is the closer the histogram approaches the true bootstrap distribution. Here is how the Monte Carlo approximation works:

1. Generate a sample with replacement from the empirical distribution for the data (this is a bootstrap sample).
2. Compute $T(F_n^*)$ the bootstrap estimate of $T(F)$. This is a replacement of the original sample with a bootstrap sample and the bootstrap estimate of $T(F)$ in place of the sample estimate of $T(F)$.
3. Repeat steps 1 and 2 M times where M is large, say 100,000.

Now a very important thing to remember is that with the Monte Carlo approximation to the bootstrap, there are two sources of error:

1. the Monte Carlo approximation to the bootstrap distribution, which can be made as small as you like by making M large;
2. the approximation of the bootstrap distribution F_n^* to the population distribution F.

If $T(F_n^*)$ converges to $T(F)$ as $n \to \infty$, then bootstrapping works. It is nice that this works out often, but it is not guaranteed. We know by a theorem called the Glivenko–Cantelli theorem that F_n converges to F uniformly. Often, we know that the sample estimate is consistent (as is the case for the sample mean). So, (1) $T(F_n)$ converges to $T(F)$ as $n \to \infty$. But this is dependent on smoothness conditions on the functional T. So we also need (2) $T(F_n^*) - T(F_n)$ to tend to 0 as $n \to \infty$. In proving that bootstrapping works (i.e., the bootstrap estimate is consistent for the population parameter), probability theorists needed to verify (1) and (2). One approach that is commonly used is by verifying that smoothness conditions are satisfied for expansions like the Edgeworth and Cornish–Fisher expansions. Then, these expansions are used to prove the limit theorems.

The probability theory associated with the bootstrap is beyond the scope of this text and can be found in books such as Hall (1992). What is important is that we know that consistency of bootstrap estimates has been demonstrated in many cases and examples where certain bootstrap estimates fail to be consistent are also known. There is a middle ground, which are cases where consistency has been neither proved nor disproved. In those cases, simulation studies can be used to confirm or deny the usefulness of the bootstrap estimate. Also, simulation studies can be used when the sample size is too small to count on asymptotic theory, and its use in small to moderate sample sizes needs to be evaluated.

1.2.1 Jackknife

The jackknife was introduced by Quenouille (1949). Quenouille's aim was to improve an estimate by correcting for its bias. Later on, Tukey (1958) popularized the method and found that a more important use of the jackknife was to estimate standard errors of an estimate. It was Tukey who coined the name jackknife because it was a statistical tool with many purposes. While bootstrapping uses the bootstrap samples to estimate variability, the jackknife uses what are called pseudovalues.

First, consider an estimate $\tilde{u}$ based on a sample of size n of observations independently drawn from a common distribution F. Here, just as with the bootstrap, we again let F_n be the empirical distribution for this data set and assume that the parameter $u = T(F)$, a functional; $\tilde{u} = T(F_n)$, and $\tilde{u}_{(i)} = T(F_{n(i)})$, where $F_{n(i)}$ is the empirical distribution function for the $n - 1$ observations obtained by leaving the ith observation out. If $\tilde{u}$ is the population variance, the jackknife estimate of variance of σ^2 is obtained as follows:

$$\sigma^2_{\text{JACK}} = n \sum_{i=1}^{n} \left(\tilde{u}_{(i)} - u^* \right)^2 \Big/ (n-1),$$

where $u^* = \sum_{i=1}^{n} \tilde{u}_{(i)} / n$. The jackknife estimate of standard error for $\tilde{u}$ is just the square root of σ^2_{JACK}. Tukey defined the pseudovalue as $\tilde{u}_i = \tilde{u} + (n-1)(\tilde{u} - \tilde{u}_{(i)})$. Then the jackknife estimate of the parameter u is $u_{\text{JACK}} = \sum_{i=1}^{n} \tilde{u}_i / n$. So the name pseudovalue comes about because the estimate is the average of the pseudovalues. Expressing the estimate of the variance of the estimate $\tilde{u}$ in terms of the pseudovalues we get

$$\sigma^2_{\text{JACK}} = \sum_{i=1}^{n} \left(\tilde{u}_i - u_{\text{JACK}} \right)^2 \Big/ [n(n-1)].$$

In this form, we see that the variance is the usual estimate for variance of a sample mean. In this case, it is the sample mean of the pseudovalues. Like the bootstrap, the jackknife has been a very useful tool in estimating variances for more complicated estimators such as trimmed or Winsorized means.

One of the great surprises about the bootstrap is that in cases like the trimmed mean, the bootstrap does better than the jackknife (Efron, 1982, pp. 28–29). For the sample median, the bootstrap provides a consistent estimate of the variance but the jackknife does not! See Efron (1982, p. 16 and chapter 6). In that monograph,

Efron also showed, using theorem 6.1, that the jackknife estimate of standard error is essentially the bootstrap estimate with the parameter estimate replaced by a linear approximation of it. In this way, there is a close similarity between the two methods, and if the linear approximation is a good approximation, the jackknife and the bootstrap will both be consistent. However, there are complex estimators where this is not the case.

1.2.2 Delta Method

It is often the case that we are interested in the moments of an estimator. In particular, for these various methods, the variance is the moment we are most interested in. To illustrate the delta method, let us define $\varphi = f(\alpha)$ where the parameters φ and α are both one-dimensional variables and f is a function differentiable with respect to α. So there exists a Taylor series expansion for f at a point say α_0. Carrying it out only to first order, we get $\varphi = f(\alpha) = f(\alpha_0) + (\alpha - \alpha_0)f'(\alpha_0)$ + remainder terms and dropping the remainder terms leaves

$$\varphi = f(\alpha) = f(\alpha_0) + (\alpha - \alpha_0) f'(\alpha_0)$$

or

$$f(\alpha) - f(\alpha_0) = (\alpha - \alpha_0) f'(\alpha_0).$$

Squaring both sides of the last equation gives us $[f(\alpha) - f(\alpha_0)]^2 = (\alpha - \alpha_0)^2[f'(\alpha_0)]^2$. Now we want to think of $\varphi = f(\alpha)$ as a random variable, and upon taking expectations of the random variables on each side of the equation, we get

$$E[f(\alpha) - f(\alpha_0)]^2 = E(\alpha - \alpha_0)^2 [f'(\alpha_0)]^2. \tag{1.1}$$

Here, α and $f(\alpha)$ are random variables, and $\alpha_0, f(\alpha_0)$, and $f'(\alpha_0)$ are all constants. Equation 1.1 provides the delta method approximation to the variance of $\varphi = f(\alpha)$ since the left-hand side is approximately the variance of φ and the right-hand side is the variance of α multiplied by the constant $[f'(\alpha_0)]^2$ if we choose α_0 to be the mean of α.

1.2.3 Cross-Validation

Cross-validation is a general procedure used in statistical modeling. It can be used to determine the best model out of alternative choices such as order of an autoregressive time series model, which variables to include in a logistic regression or a multiple linear regression, number of distributions in a mixture model, and the choice of a parametric classification model or for pruning classification trees.

The basic idea of cross-validation is to randomly split the data into two subsets. One is used to fit the model, and the other is used to test the model. The extreme case would be to fit all the data except for a single observation and see how well that model predicts the value of the observation left out. But a sample of size 1 is not very good for assessment. So, in the case of classification error rate estimation, Lachenbruch and Mickey (1968) proposed the leave-one-out method of assessment. In this case, a model is fit to the $n - 1$ observations that are included and is tested on the one left out. But the model fitting and prediction is then done separately for all n observations by testing the model fit without observation i for predicting the class for the case i. Results are obtained from each i and then averaged. Efron (1983) included a simulation study that showed for bivariate normal distributions the "632" variant of the bootstrap does better than leave-one-out. For pruning classification trees, see Brieman et al. (1984).

1.2.4   Subsampling

The idea of subsampling goes back to Hartigan (1969), who developed a theory of confidence intervals for random subsampling. He proved a theorem called the typical value theorem when M-estimators are used to estimate parameters. We shall see in the chapter on confidence intervals that Hartigan's results were motivating factors for Efron to introduce the percentile method bootstrap confidence intervals.

More recently the theory of subsampling has been further developed and related to the bootstrap. It has been applied when the data are independent observations and also when there are dependencies among the data. A good summary of the current literature along with connections to the bootstrap can be found in Politis et al. (1999), and consistency under very minimal assumptions can be found in Politis and Romano (1994). Politis, Romano, and Wolf included applications when the observations are independent and also for dependent situations such as stationary and nonstationary time series, random fields, and marked point processes. The dependent situations are also well covered in section 2.8 of Lahiri (2003).

We shall now define random subsampling. Let $S_1, S_2, \ldots, S_{B-1}$ be $B - 1$ of the $2^n - 1$ nonempty subsets of the integers $1, 2, \ldots, n$. These $B - 1$ subsets are selected at random without replacement. So a subset of size 3 might be drawn, and it would contain $\{1, 3, 5\}$. Another subset of size 3 that could be drawn could be $\{2, 4, n\}$. Subsets of other sizes could also be drawn. For example, a subset of size 5 is $\{1, 7, 9, 12, 13\}$. There are many subsets to select from. There is only 1 subset of size n, and it contains all the integers from 1 to n. There are n subsets of size $n - 1$. Each distinct subset excludes one and only one of the integers from 1 to n. For more details on this and M-estimators and the typical value theorem see sections 3.1.1 and 3.1.2 of Chernick (2007).

1.3   WIDE RANGE OF APPLICATIONS

There is a great deal of temptation to apply the bootstrap in a wide variety of settings. But as we have seen, the bootstrap does not always work. So how do we know when it will work? We either have to prove a consistency theorem under a set of assumptions or we have to verify that it is well behaved through simulations.

In regression problems, there are at least two approaches to bootstrapping. One is called "bootstrapping residuals," and the other is called "bootstrapping vectors or cases." In the first approach, we fit a model to the data and compute the residuals from the model. Then we generate a bootstrap sample by resampling with replacement from the model residuals. In the second approach, we resample with replacement from the n, $k + 1$ dimensional vectors:

$$(y_i, X_{1i}, X_{2i}, \ldots, X_{ki}) \quad \text{for} \quad I = 1, 2, \ldots, n.$$

In the first approach, the model is fixed. In the second, it is redetermined each time. Both methods can be applied when a parametric regression model is assumed. But in practice, we might not be sure that the parametric form is correct. In such cases, it is better to use the bootstrapping vectors approach.

The bootstrap has also been successfully applied to the estimation of error rates for discriminant functions using bias adjustment as we will see in Chapter 2. The bootstrap and another resampling procedure called "permutation tests," as described in Good (1994), are attractive because they free the scientists from restrictive parametric assumptions that may not apply in their particular situation.

Sometimes the data can have highly skewed or heavy-tailed distributions or multiple modes. There is no need to simplify the model by, say, a linear approximation when the appropriate model is nonlinear. The estimator can be defined through an algorithm and there does not need to be an analytic expression for the parameters to be estimated.

Another feature of the bootstrap is its simplicity. For almost any problem you can think of, there is a way to construct bootstrap samples. Using the Monte Carlo approximation to the bootstrap estimate, all the work can be done by the computer. Even though it is a computer-intensive method, with the speed of the modern computer, most problems are feasible, and in many cases, up to 100,000 bootstrap samples can be generated without consuming hours of CPU time. But care must be taken. It is not always apparent when the bootstrap will fail, and failure may not be easy to diagnose.

In recent years, we are finding that there are ways to modify the bootstrap so that it will work for problems where the simple (or naïve) bootstrap is known to fail. The "m-out-n" bootstrap is one such example.

In many situations, the bootstrap can alert the practitioner to variability in his procedures that he otherwise would not be aware of. One example in spatial statistics is the development of pollution level contours based on a smoothing method called "kriging." By generating bootstrap samples, multiple kriging contour maps can be generated, and the differences in the contours can be determined visually.

Also, the stepwise logistic regression problem that is described in Gong (1986) shows that variable selection can be somewhat of a chance outcome when there are many competing variables. She showed this by bootstrapping the entire stepwise selection procedure and seeing that the number of variables and the choice of variables selected can vary from one bootstrap sample to the next.

Babu and Feigelson (1996) applied the bootstrap to astronomy problems. In clinical trials, the bootstrap is used to estimate individual bioequivalence, for P-value adjustment with multiple end points, and even to estimate mean differences when the sample size is not large enough for asymptotic theory to take hold or the data are very non-normal and statistics other that the mean are important.

1.4   THE BOOTSTRAP AND THE R LANGUAGE SYSTEM

In subsequent chapters of this text, we will illustrate examples with calculations and short programs using the R language system and its associated packages.

R is an integrated suite of an object-oriented programming language and software facilities for data manipulation, calculation, and graphical display. Over the last decade, R has become the statistical environment of choice for academics, and probably is now the most used such software system in the world. The number of specialized packages available in R has increased exponentially, and continues to do so. Perhaps the best thing about R (besides its power and breadth) is this: It is completely free to use. You can obtain your own copy of the R system at http://www.cran.r-project.org/.

From this website, you can get not only the executable version of R for Linux, Macs, or Windows, but also even the source programs and free books containing documentation. We have found *The R Book* by Michael J. Crawley a good way to learn how to use R, and have found it to be an invaluable reference afterword.

There are so many good books and courses from which you can learn R, including courses that are Internet based, such as at http://statistics.com. We will not attempt to teach even the basics of R here. What we will do is show those features of direct applicability, and give program snippets to illustrate examples and the use of currently available R packages for bootstrapping. These snippets will be presented in the Courier typeface to distinguish them from regular text and to maintain spacing in output generated.

At the current time, using R version 2.10.1, the R query (">" denotes the R command line prompt)

```
> ?? bootstrap

or

> help.search('bootstrap')

results in

agce::resamp.std Compute the standard
deviation by bootstrap.

alr3::boot.case Case bootstrap for
regression models

analogue::RMSEP Root mean square error of
prediction

analogue::bootstrap Bootstrap estimation and
errors

analogue::bootstrap.waBootstrap estimation and
errors for WA models

analogue::bootstrapObject Bootstrap object
description
```

analogue::getK Extract and set the number of analogues

analogue::performance Transfer function model performance statistics

analogue::screeplot.mat Screeplots of model results

analogue::summary.bootstrap.mat Summarise bootstrap resampling for MAT models

animation::boot.iid Bootstrapping the i.i.d data

ape::boot.phylo Tree Bipartition and Bootstrapping Phylogenies

aplpack::slider.bootstrap.lm.plot interactive bootstapping for lm

bnlearn::bn.boot Parametric and nonparametric bootstrap of Bayesian networks

bnlearn::boot.strength Bootstrap arc strength and direction

boot::nested.corr Functions for Bootstrap Practicals

boot::boot Bootstrap Resampling

boot::boot.array Bootstrap Resampling Arrays

boot::boot.ci Nonparametric Bootstrap Confidence Intervals

boot::cd4.nested Nested Bootstrap of cd4 data

boot::censboot Bootstrap for Censored Data

boot::freq.array Bootstrap Frequency Arrays

boot::jack.after.boot Jackknife-after-Bootstrap Plots

boot::linear.approx Linear Approximation of Bootstrap Replicates

boot::plot.boot Plots of the Output of a Bootstrap Simulation

boot::print.boot Print a Summary of a Bootstrap Object

boot::print.bootci Print Bootstrap Confidence Intervals

boot::saddle Saddlepoint Approximations for Bootstrap Statistics

boot::saddle.distn Saddlepoint Distribution Approximations for Bootstrap Statistics boot::tilt.boot Non-parametric Tilted Bootstrap

boot::tsboot Bootstrapping of Time Series

BootCL::BootCL.distribution Find the bootstrap distribution

BootCL::BootCL.plot Display the bootstrap distribution and p-value

BootPR::BootAfterBootPI Bootstrap-after-Bootstrap Prediction

BootPR::BootBC Bootstrap bias-corrected estimation and forecasting for AR models

BootPR::BootPI Bootstrap prediction intevals and point forecasts with no bias-correction

BootPR::BootPR-package Bootstrap Prediction Intervals and Bias-Corrected Forecasting

BootPR::ShamanStine.PI Bootstrap prediction interval using Shaman and Stine bias formula

bootRes::bootRes-package The bootRes Package for Bootstrapped Response and Correlation Functions

bootRes::dendroclim Calculation of bootstrapped response and correlation functions.

bootspecdens::specdens Bootstrap for testing equality of spectral densities

bootStepAIC::boot.stepAIC Bootstraps the Stepwise Algorithm of stepAIC() for Choosing a Model by AIC

bootstrap::bootpred Bootstrap Estimates of Prediction Error

bootstrap::bootstrap Non-Parametric Bootstrapping

bootstrap::boott Bootstrap-t Confidence Limits

bootstrap::ctsub Internal functions of package bootstrap

bootstrap::lutenhorm Luteinizing Hormone

bootstrap::scor Open/Closed Book Examination Data

bootstrap::spatial Spatial Test Data

BSagri::BOOTSimpsonD Simultaneous confidence intervals for Simpson indices

cfa::bcfa Bootstrap-CFA

ChainLadder::BootChainLadder Bootstrap-Chain-Ladder Model

CircStats::vm.bootstrap.ci Bootstrap Confidence Intervals

circular::mle.vonmises.bootstrap.ci Bootstrap Confidence Intervals

clue::cl_boot Bootstrap Resampling of Clustering Algorithms

CORREP::cor.bootci Bootstrap Confidence Interval for Multivariate Correlation

Daim::Daim.data1 Data set: Artificial bootstrap data for use with Daim

DCluster::achisq.boot Bootstrap replicates of Pearson's Chi-square statistic

DCluster::besagnewell.boot Generate boostrap replicates of Besag and Newell's statistic

DCluster::gearyc.boot Generate bootstrap replicates of Moran's I autocorrelation statistic

DCluster::kullnagar.boot Generate bootstrap replicates of Kulldorff and Nagarwalla's statistic

DCluster::moranI.boot Generate bootstrap replicates of Moran's I autocorrelation statistic

DCluster::pottwhitt.boot Bootstrap replicates of Potthoff-Whittinghill's statistic

DCluster::stone.boot Generate boostrap replicates of Stone's statistic

DCluster::tango.boot Generate bootstrap replicated of Tango's statistic

DCluster::whittermore.boot Generate bootstrap replicates of Whittermore's statistic

degreenet::rplnmle Rounded Poisson Lognormal Modeling of Discrete Data

degreenet::bsdp Calculate Bootstrap Estimates and Confidence Intervals for the Discrete Pareto Distribution

degreenet::bsnb Calculate Bootstrap Estimates and Confidence Intervals for the Negative Binomial Distribution

degreenet::bspln Calculate Bootstrap Estimates and Confidence Intervals for the Poisson Lognormal Distribution

degreenet::bswar Calculate Bootstrap Estimates and Confidence Intervals for the Waring Distribution degreenet::bsyule Calculate Bootstrap Estimates and Confidence Intervals for the Yule Distribution

degreenet::degreenet-internal Internal degreenet Objects

delt::eval.bagg Returns a bootstrap aggregation of adaptive histograms

delt::lstseq.bagg Calculates a scale of bootstrap aggregated histograms

depmix::depmix Fitting Dependent Mixture Models

Design::anova.Design Analysis of Variance (Wald and F Statistics)

Design::bootcov Bootstrap Covariance and Distribution for Regression Coefficients

Design::calibrate Resampling Model Calibration

Design::predab.resample Predictive Ability using Resampling

Design::rm.impute Imputation of Repeated Measures

Design::validate Resampling Validation of a Fitted Model's Indexes of Fit

Design::validate.cph Validation of a Fitted Cox or Parametric Survival Model's Indexes of Fit

Design::validate.lrm Resampling Validation of a Logistic Model

Design::validate.ols Validation of an Ordinary Linear Model

dynCorr::bootstrapCI Bootstrap Confidence Interval

dynCorr::dynCorrData An example dataset for use in the example calls in the help files for the dynamicCorrelation and bootstrapCI functions

e1071::bootstrap.lca Bootstrap Samples of LCA Results

eba::boot Bootstrap for Elimination-By-Aspects (EBA) Models

EffectiveDose::Boot.CI Bootstrap confidence intervals for ED levels

EffectiveDose::EffectiveDose-package Estimation of the Effective Dose including Bootstrap confindence intervals

el.convex::samp sample from bootstrap

equate::se.boot Bootstrap Standard Errors of Equating

equivalence::equiv.boot Regression-based TOST using bootstrap

extRemes::boot.sequence Bootstrap a sequence.

FactoMineR::simule Simulate by bootstrap

FGN::Boot Generic Bootstrap Function

FitAR::Boot Generic Bootstrap Function

FitAR::Boot.ts Parametric Time Series Bootstrap

fitdistrplus::bootdist Bootstrap simulation of uncertainty for non-censored data

fitdistrplus::bootdistcens Bootstrap simulation of uncertainty for censored data

flexclust::bootFlexclust Bootstrap Flexclust Algorithms

fossil::bootstrap Bootstrap Species Richness Estimator

fractal::surrogate Surrogate data generation

FRB::FRBmultiregGS GS-Estimates for multivariate regression with bootstrap confidence intervals

FRB::FRBmultiregMM MM-Estimates for Multivariate Regression with Bootstrap Inference

FRB::FRBmultiregS S-Estimates for Multivariate Regression with Bootstrap Inference

FRB::FRBpcaMM PCA based on Multivariate MM-estimators with Fast and Robust Bootstrap

FRB::FRBpcaS PCA based on Multivariate S-estimators with Fast and Robust Bootstrap

FRB::GSboot_multireg Fast and Robust Bootstrap for GS-Estimates

FRB::MMboot_loccov Fast and Robust Bootstrap for MM-estimates of Location and Covariance

FRB::MMboot_multireg Fast and Robust Bootstrap for MM-Estimates of Multivariate Regression

FRB::MMboot_twosample Fast and Robust Bootstrap for Two-Sample MM-estimates of Location and Covariance

FRB::Sboot_loccov Fast and Robust Bootstrap for S-estimates of location/covariance

FRB::Sboot_multireg Fast and Robust Bootstrap for S-Estimates of Multivariate Regression

FRB::Sboot_twosample Fast and Robust Bootstrap for Two-Sample S-estimates of Location and Covariance

ftsa::fbootstrap Bootstrap independent and identically distributed functional data

gmvalid::gm.boot.coco Graphical model validation using the bootstrap (CoCo).

gmvalid::gm.boot.mim Graphical model validation using the bootstrap (MIM)

gPdtest::gPd.test Bootstrap goodness-of-fit test for the generalized Pareto distribution

hierfstat::boot.vc Bootstrap confidence intervals for variance components

Hmisc::areg Additive Regression with Optimal Transformations on Both Sides using Canonical Variates

Hmisc::aregImpute Multiple Imputation using Additive Regression, Bootstrapping, and Predictive Mean Matching

Hmisc::bootkm Bootstrap Kaplan-Meier Estimates

Hmisc::find.matches Find Close Matches

Hmisc::rm.boot Bootstrap Repeated Measurements Model

Hmisc::smean.cl.normal Compute Summary Statistics on a Vector

Hmisc::transace Additive Regression and Transformations using ace or avas

Hmisc::transcan Transformations/Imputations using Canonical Variates

homtest::HOMTESTS Homogeneity tests

hopach::boot2fuzzy function to write MapleTree files for viewing bootstrap estimated cluster membership probabilities based on hopach clustering results

hopach::bootplot function to make a barplot of bootstrap estimated cluster membership probabilities

hopach::boothopach functions to perform non-parametric bootstrap resampling of hopach clustering results

ICEinfer::ICEcolor Compute Preference Colors for Outcomes in a Bootstrap ICE Scatter within a Confidence Wedge

ICEinfer::ICEuncrt Compute Bootstrap Distribution of ICE Uncertainty for given Shadow Price of Health, lambda

ICEinfer::plot.ICEcolor Add Economic Preference Colors to Bootstrap Uncertainty Scatters within a Confidence Wedge

ICEinfer::plot.ICEuncrt Display Scatter for a possibly Transformed Bootstrap Distribution of ICE Uncertainty

ICEinfer::print.ICEuncrt Summary Statistics for a possibly Transformed Bootstrap Distribution of ICE Uncertainty

ipred::bootest Bootstrap Error Rate Estimators

maanova::consensus Build consensus tree out of bootstrap cluster result

Matching::ks.boot Bootstrap Kolmogorov-Smirnov

MBESS::ci.reliability.bs Bootstrap the confidence interval for reliability coefficient

MCE::RProj The bootstrap-then-group implementation of the Bootstrap Grouping Prediction Plot for estimating R.

MCE::groupbootMCE The group-then-bootstrap implementation of the Bootstrap Grouping Prediction Plot for estimating MCE

MCE::groupbootR The group-then-bootstrap implementation of the Bootstrap Grouping Prediction Plot for estimating R

MCE::jackafterboot Jackknife-After-Bootstrap Method of MCE estimation

MCE::mceBoot Bootstrap-After-Bootstrap estimate of MCE

MCE::mceProj The bootstrap-then-group implementation of the Bootstrap Grouping Prediction Plot for estimating MCE.

meboot::meboot Generate Maximum Entropy Bootstrapped Time Series Ensemble

meboot::meboot.default Generate Maximum Entropy Bootstrapped Time Series Ensemble

meboot::meboot.pdata.frame Maximum Entropy Bootstrap for Panel Time Series Data meifly::lmboot Bootstrap linear models

mixreg::bootcomp Perform a bootstrap test for the number of components in a mixture of regressions.

mixstock::genboot Generate bootstrap estimates of mixed stock analyses

mixstock::mixstock.boot Bootstrap samples of mixed stock analysis data

mixtools::boot.comp Performs Parametric Bootstrap for Sequentially Testing the Number of Components in Various Mixture Models

mixtools::boot.se Performs Parametric Bootstrap for Standard Error Approximation

MLDS::simu.6pt Perform Bootstrap Test on 6-point Likelihood for MLDS FIT

MLDS::summary.mlds.bt Method to Extract Bootstrap Values for MLDS Scale Values

msm::boot.msm Bootstrap resampling for multi-state models

mstate::msboot Bootstrap function in multi-state models

multtest::boot.null Non-parametric bootstrap resampling function in package 'multtest'

ncf::mSynch the mean (cross-)correlation (with bootstrap CI) for a panel of spatiotemporal data

nFactors::eigenBootParallel Bootstrapping of the Eigenvalues From a Data Frame

nlstools::nlsBoot Bootstrap resampling

np::b.star Compute Optimal Block Length for Stationary and Circular Bootstrap

nsRFA::HOMTESTS Homogeneity tests

Oncotree::bootstrap.oncotree Bootstrap an oncogenetic tree to assess stability

ouch::browntree Fitted phylogenetic Brownian motion model

ouch::hansentree-methods Methods of the "hansentree" class

pARccs::Boot_CI Bootstrap confidence intervals for (partial) attributable risks (AR and PAR) from case-control data

PCS::PdCSGt.bootstrap.NP2 Non-parametric bootstrap for computing G-best and d-best PCS

PCS::PdofCSGt.bootstrap5 Parametric bootstrap for computing G-best and d-best PCS

PCS::PofCSLt.bootstrap5 Parametric bootstrap for computing L-best PCS

peperr::complexity.ipec.CoxBoost Interface function for complexity selection for CoxBoost via integrated prediction error curve and the bootstrap

peperr::complexity.ipec.rsf_mtry Interface function for complexity selection for random survival forest via integrated prediction error curve and the bootstrap

pgirmess::difshannonbio Empirical confidence interval of the bootstrap of the difference between two Shannon indices

pgirmess::piankabioboot Bootstrap Pianka's index

pgirmess::shannonbioboot Boostrap Shannon's and equitability indices

phangorn::bootstrap.pml Bootstrap

phybase::bootstrap Bootstrap sequences

phybase::bootstrap.mulgene Bootstrap sequences from multiple loci

popbio::boot.transitions Bootstrap observed census transitions

popbio::countCDFxt Count-based extinction probabilities and bootstrap confidence intervals

prabclus::abundtest Parametric bootstrap test for clustering in abundance matrices

prabclus::prabtest Parametric bootstrap test for clustering in presence-absence matrices

pvclust::msfit Curve Fitting for Multiscale Bootstrap Resampling

qgen::dis Bootstrap confidence intervals

qpcR::calib2 Calculation of qPCR efficiency by dilution curve analysis and bootstrapping of dilution curve replicates

qpcR::pcrboot Bootstrapping and jackknifing qPCR data

qtl::plot.scanoneboot Plot results of bootstrap for QTL position

qtl::scanoneboot Bootstrap to get interval estimate of QTL location qtl::summary.scanoneboot Bootstrap confidence interval for QTL location

QuantPsyc::distInd.ef Complex Mediation for use in Bootstrapping

QuantPsyc::proxInd.ef Simple Mediation for use in Bootstrapping

quantreg::boot.crq Bootstrapping Censored Quantile Regression

quantreg::boot.rq Bootstrapping Quantile Regression

r4ss::SS_splitdat Split apart bootstrap data to make input file.

relaimpo::boot.relimp Functions to Bootstrap Relative Importance Metrics

ResearchMethods::bootSequence A demonstration of how bootstrapping works, taking multiple bootstrap samples and watching how the means of those samples begin to normalize.

ResearchMethods::bootSingle A demonstration of how bootstrapping works step by step for one function.

rms::anova.rms Analysis of Variance (Wald and F Statistics)

rms::bootcov Bootstrap Covariance and Distribution for Regression Coefficients

rms::calibrate Resampling Model Calibration

rms::predab.resample Predictive Ability using Resampling

rms::validate Resampling Validation of a Fitted Model's Indexes of Fit

rms::validate.cph Validation of a Fitted Cox or Parametric Survival Model's Indexes of Fit

rms::validate.lrm Resampling Validation of a Logistic Model

rms::validate.ols Validation of an Ordinary Linear Model

robust::rb Robust Bootstrap Standard Errors

rqmcmb2::rqmcmb Markov Chain Marginal Bootstrap for Quantile Regression

sac::BootsChapt Bootstrap (Permutation) Test of Change-Point(s) with One-Change or Epidemic Alternative

sac::BootsModelTest Bootstrap Test of the Validity of the Semiparametric Change-Point Model

SAFD::btest.mean One-sample bootstrap test for the mean of a FRV

SAFD::btest2.mean Two-sample bootstrap test on the equality of mean of two FRVs

SAFD::btestk.mean Multi-sample bootstrap test for the equality of the mean of FRVs

scaleboot::sboptions Options for Multiscale Bootstrap

scaleboot::plot.scaleboot Plot Diagnostics for Multiscale Bootstrap

scaleboot::sbconf Bootstrap Confidence Intervals

scaleboot::sbfit Fitting Models to Bootstrap Probabilities

scaleboot::scaleboot-package Approximately Unbiased P-values via Multiscale Bootstrap

scaleboot::scaleboot Multiscale Bootstrap Resampling

scaleboot::summary.scaleboot P-value Calculation for Multiscale Bootstrap

sem::boot.sem Bootstrap a Structural Equation Model

shapes::resampletest Tests for mean shape difference using complex arithmetic, including bootstrap and permutation tests.

shapes::iglogl Internal function(s)

shapes::testmeanshapes Tests for mean shape difference, including permutation and bootstrap tests

simpleboot::hist.simpleboot Histograms for bootstrap sampling distributions.

simpleboot::lm.boot Linear model bootstrap.

simpleboot::summary.lm.simpleboot Methods for linear model bootstrap.

simpleboot::loess.boot 2-D Loess bootstrap.

simpleboot::fitted.loess.simpleboot Methods for loess bootstrap.

simpleboot::one.boot One sample bootstrap of a univariate statistic.

simpleboot::pairs.boot Two sample bootstrap.

simpleboot::perc Extract percentiles from a bootstrap sampling distribution.

simpleboot::plot.lm.simpleboot Plot method for linear model bootstraps.

simpleboot::plot.loess.simpleboot Plot method for loess bootstraps.

simpleboot::samples Extract sampling distributions from bootstrapped linear/loess models.

simpleboot::two.boot Two sample bootstrap of differences between univariate statistics.

sm::sm.binomial.bootstrap Bootstrap goodness-of-fit test for a logistic regression model.

sm::sm.poisson.bootstrap Bootstrap goodness-of-fit test for a Poisson regression model

spls::ci.spls Calculate bootstrapped confidence intervals of SPLS coefficients

spls::correct.spls Correct the initial SPLS coefficient estimates based on bootstrapped confidence intervals

Stem::covariates2 Stem internal objects

Stem::Stem.Bootstrap Parametric bootstrap

survey::bootweights Compute survey bootstrap weights

tractor.base::angleBetweenVectors Undocumented functions

TSA::arima.boot Compute the Bootstrap Estimates of an ARIMA Model

tsDyn::TVAR.sim Simulation and bootstrap of multivariate Threshold Autoregressive model

tsDyn::TVECM.sim Simulation and bootstrap of bivariate VECM/TVECM

tsDyn::extendBoot extension of the bootstrap replications

tsDyn::setar.sim Simulation and bootstrap of Threshold Autoregressive model

tseries::tsbootstrap Bootstrap for General Stationary Data

ttrTests::bootstrap Generates a Bootstrap Sample from Raw Data

ttrTests::generateSample Generates a Bootstrap Sample from Price Data

TWIX::bootTWIX Bootstrap of the TWIX trees

UsingR::cfb Bootstrap sample from the Survey of Consumer Finances

varSelRF::varSelRFBoot Bootstrap the variable selection procedure in varSelRF

```
    vegetarian::bootstrap Estimates
Uncertainties with Bootstrapping

    verification::table.stats.boot Percentile
bootstrap for 2 by 2 table

    vrtest::AutoBoot.test Wild Bootstrapping of
Automatic Variance Ratio Test

    vrtest::Boot.test Bootstrap Variance Ratio
Tests

    waveslim::dwpt.boot Bootstrap Time Series
Using the DWPT

    wmtsa::wavBootstrap Adaptive wavelet-based
bootstrapping

    wmtsa::wavDWPTWhitest Seeks the whitest
transform of a discrete wavelet packet transform
(DWPT)
```

The part of the name before ":::" is the "package" name (class), which installs a library that has the "function" object whose name follows the ":::." The above list should indicate both the breadth of applications of the bootstrap and the breadth of its implementation in the R system.

R comes with some basic packages preinstalled. Most special application packages have to be downloaded by the user via the menu line command Packages | Install Packages. This makes the chosen packages(s) part of the R software on your computer. To actually bring the package into use in your environment, you will also need the require() or library() functions. Two packages of note related to bootstrapping are the "bootstrap" package, which is documented by the book *An Introduction to the Bootstrap* by B. Efron and R. J. Tibshirani, and the "boot" package, which is documented by *Bootstrap Methods and Their Application* by A. C. Davison and D. V. Hinkley. For example, you can require the "boot" library by

```
> require('boot')
Loading required package: boot
```

R is a vectorized and object-oriented language. Most operations are most efficient when done as vector operations instead of on individual elements. For example,

```
> x<- 1:10
> y<- 21:30
> x
[1] 1 2 3 4 5 6 7 8 9 10
> y
[1] 21 22 23 24 25 26 27 28 29 30
> x+y
[1] 22 24 26 28 30 32 34 36 38 40
```

```
> x/y
```
```
[1]  0.04761905 0.09090909 0.13043478
0.16666667 0.20000000 0.23076923 0.25925926
0.28571429 0.31034483 0.33333333
```
```
> x*y
```
```
[1] 21 44 69 96 125 156 189 224 261 300
```
```
> sqrt(x)
```
```
[1]  1.000000 1.414214 1.732051 2.000000
2.236068 2.449490 2.645751 2.828427 3.000000
3.162278
```
```
> exp(x)
```
```
[1]  2.718282 7.389056 20.085537 54.598150
148.413159 403.428793 1096.633158 2980.957987
8103.083928
```
```
[10]  22026.465795
```
```
> x[2]
```
```
[1]  2
```
```
> x[2]+y[3]
```
```
[1]  25
```

Note that individual elements are indicated by subscripts within brackets "[]," and "n:m" is shorthand for the vector whose elements are the sequence of integers from n to m.

One function in the basic R packages that lies at the heart of resampling is the sample() function, whose syntax is

```
sample(x, size, replace = FALSE, prob = NULL)
```

The first argument "x" is the vector of data, that is, the original sample. "size" is the size of the resample desired. "replace" is "TRUE" if resampling is with replacement, and "FALSE" if not (the default). "prob" is a vector of probability weights if the equal-weight default is not used. Any arguments omitted will assume the default. If "size" is omitted, it will default to the length of "x."

For our purposes, it will usually be easiest to resample the indices of the data from a sample of size n, rather than the data itself. For example, if we have five data in our set, say

```
> x<- c(-0.3, 0.5, 2.6, 1.0, -0.9)
```
```
> x
```
```
[1]  -0.3 0.5 2.6 1.0 -0.9
```
```
then
```
```
> i<- sample(1:5, 5, replace=TRUE)
```
```
> i
```
```
[1]  3 2 3 2 2
```

```
> x[i]
[1] 2.6 0.5 2.6 0.5 0.5
```

is the resample of the original data.

As we move through the text, more features of R related to the bootstrap will be illustrated in context, as we need them.

1.5 HISTORICAL NOTES

Bootstrap research began in earnest in the late 1970s, although some key developments can be traced back to earlier times. The theory took off in the early 1980s after Efron (1979). The first proofs of the consistency of the bootstrap estimate of the sample mean came in 1981 with the papers of Singh (1981) and Bickel and Freedman (1981).

The significance of Efron (1979) is best expressed in Davison and Hinkley (1997) who wrote " The publication in 1979 of Bradley Efron's first article on bootstrap methods was a major event in Statistics, at once synthesizing some of the earlier resampling ideas and establishing a new framework for simulation-based statistical analysis. The idea of replacing complicated and often inaccurate approximations to biases, variances and other measures of uncertainty by computer simulation caught the imagination of both theoretical researchers and users of statistical methods."

Regarding the precursors of the bootstrap, Efron pointed out some of the early work of R. A. Fisher (in the 1920s on maximum likelihood estimation) as the inspiration for many of the basic ideas. The jackknife was introduced by Quenouille (1949) and popularized by Tukey (1958). Miller (1974) provided an excellent review of jackknife methods. Extensive coverage of the jackknife as developed up to 1972 can be found in Gray and Schucany (1972).

As noted earlier, Bickel and Freedman (1981) and Singh (1981) were the first to show consistency of the bootstrap estimate of the sample mean under certain regularity conditions. In their paper, Bickel and Freedman (1981) also provided an example where the bootstrap estimate is not consistent. Gine and Zinn (1989) provided necessary but not sufficient conditions for the consistency of the bootstrap mean.

Athreya (1987a, b), Knight (1989), and Angus (1993) all provided examples where the bootstrap fails due to the fact that the necessary conditions were not satisfied. In some of these cases the inconsistency is shown by deriving the actual limiting distribution for the bootstrap estimator normalized and by showing that it is not degenerate but differs from the limiting distribution for the original parameter estimate.

Subsampling methods began with Hartigan (1969, 1971, 1975) and McCarthy (1969). Diaconis and Holmes (1994) showed how the Monte Carlo approximation can sometimes be avoided through the use of Gray codes.

Efron (1983) compared several variations with the bootstrap estimate when estimating classification error rates for linear discriminant functions. Other papers that showed through simulation the advantage of the bootstrap 632 estimate include Chernick et al. (1985, 1986, 1988a, b). For the Pearson VII family, the 632 is not always the best bootstrap estimator when the parameter controlling the tail behavior increases and the first moment no longer exists (Chernick et al., 1988b). Other related papers include Chatterjee and Chatterjee (1983), McLachlan (1980), Snapinn and Knoke (1984, 1985a, b, 1988), Jain et al. (1987), and Efron and Tibshirani (1997). Many other references for the historical development of the bootstrap can be found in Chernick (2007) and Chernick and LaBudde (2010).

1.6 EXERCISES

1. Suppose three mice who are littermates have weights 82, 107, and 93 g.
    (a) What is the mean weight of the mice?
    (b) How many possible bootstrap samples of this sample are there?
    (c) List all of the possible bootstrap samples as triples.
    (d) Compute the mean of each bootstrap sample.
    (e) Compute the mean of the resample means. How does this compare with the original sample mean?
    (f) What are the high and low values of the resample means?

2. Suppose three mice in Exercise 1.6.1 have maze transit times of 27, 36, and 22 s.
    (a) What is the mean transit time of the mice?
    (b) How many possible bootstrap samples of this sample are there?
    (c) List all of the possible bootstrap samples as triples.
    (d) Compute the mean of each bootstrap resample.
    (e) Compute the mean of the resample means. How does this compare with the original sample mean?
    (f) What are the high and low values of the resample means?

3. Install the R system on your computer, and install the package "bootstrap." Then enter the following commands at the prompt:

    ```
    >require('bootstrap')
    >help('bootstrap')
    ```

    Note the webpage that appears in your browser.

4. Install the R system on your computer, and install the package "boot." Then enter the following commands at the prompt:

    ```
    >require('boot')
    >help('boot')
    ```

    Note the webpage that appears in your browser.

5. Aflatoxin residues in peanut butter: In actual testing, 12 lots of peanut butter had aflatoxin residues in parts per billion of 4.94, 5.06, 4.53, 5.07, 4.99, 5.16, 4.38, 4.43, 4.93, 4.72, 4.92, and 4.96.
    (a) How many possible bootstrap resamples of these data are there?
    (b) Using R and the sample() function, or a random number table or generator, generate five resamples of the integers from 1 to 12.
    (c) For each of the resamples in Exercise 1.6.5b, find the mean of the corresponding elements of the aflatoxin data set.

(d) Find the mean of the resample means. Compare this with the mean of the original data set.

(e) Find the minimum and the maximum of the five resample means. This a crude bootstrap confidence interval on the mean. (If you had used 1000 resamples, and used the 25th and 975th largest means, this would have given a reasonable 95% confidence interval.)

6. Sharpness of a razor blade: In a filament cut test, a razor blade was tested six different times with ultimate forces corresponding to 8.5, 13.9, 7.4, 10.3, 15.7, and 4.0 g.

(a) How many possible bootstrap resamples of these data are there?

(b) Using R and the sample() function, or a random number table or generator, generate 10 resamples of the integers from 1 to 6.

(c) For each of the resamples in Exercise 1.6.6b, find the mean of the corresponding elements of the sharpness data set.

(d) Find the mean of the resample means. Compare this with the mean of the original data set.

(e) Find the minimum and maximum of the 10 resample means. This is a crude bootstrap confidence interval on the mean. (If you had used 1000 resamples, and used the 25th and 975th largest means, this would have given a reasonable 95% confidence interval.)

REFERENCES

Angus, J. E. (1993). Asymptotic theory for bootstrapping extremes. Commun. Stat. Theory Methods 22, 15–30.

Arcones, M. A., and Gine, E. (1989). The bootstrap of the mean with arbitrary bootstrap sample size. Ann. Inst. Henri Poincare 25, 457–481.

Athreya, K. B. (1987a). Bootstrap of the mean in the infinite variance case. In Proceedings of the First World Congress of the Bernoulli Society (Y. Prohorov, and V. Sazonov, eds.), Vol. 2, pp. 95–98. VNU Science Press, The Netherlands.

Athreya, K. B. (1987b). Bootstrap estimate of the mean in the infinite variance case. Ann. Stat. 15, 724–731.

Babu, G. J., and Feigelson, E. (1996). Astrostatistics. Chapman & Hall, New York.

Beran, R. J. (1982). Estimated sampling distributions: The bootstrap and competitors. Ann. Stat. 10, 212–225.

Bickel, P. J., and Freedman, D. A. (1981). Some asymptotic theory for the bootstrap. Ann. Stat. 9, 1196–1217.

Bickel, P. J., and Freedman, D. A. (1984). Asymptotic normality and the bootstrap in stratified sampling. Ann. Stat. 12, 470–482.

Bickel, P. J., and Ren, J.-J. (1996). The m out of n bootstrap and goodness fit of tests with populations. Comput. Math. Appl. 15, 29–37.

Bickel, P. J., Goetze, F., and van Zwet, W. R. (1997). Resampling fewer than n observations, gains, losses, and remedies for losses. Stat. Sin. 7, 1–32.

Brieman, L., Friedman, J. H., Olshen, R. A., and Stone, C. J. (1984). Classification and Regression Trees. Wadsworth, Belmont.

Chatterjee, S., and Chatterjee S. (1983). Estimation of misclassification probabilities by bootstrap methods. Commun. Statist. Simul. Comput. 11, 645–656.

Chernick, M. R. (2007). Bootstrap Methods: A Guide for Practitioners and Researchers, Second Edition, Wiley, Hoboken.

Chernick, M. R., Murthy V. K., and Nealy, C. D. (1985). Applications of bootstrap and other resampling techniques: Evaluation of classifier performance. Pattern Recog. Lett. 3, 167–178.

Chernick, M. R., Murthy, V. K., and Nealy, C. D. (1986). Correction note to applications of bootstrap and other resampling techniques: Evaluation of classifier performance. Pattern Recogn. Lett. 4, 133–142.

Chernick, M. R., Murthy, V. K., and Nealy, C. D. (1988a). Estimation of error rate for linear discriminant functions by resampling: non-Gaussian populations. Comput. Math. Applic. 15, 29–37.

Chernick, M. R., Murthy, V. K., and Nealy, C. D. (1988b). Resampling-type error rate estimation for linear discriminant functions: Pearson VII distributions. Comput. Math. Appl. 15, 897–902.

Chernick, M. R. and LaBudde, R. A. (2010). Revisiting qualms about bootstrap confidence intervals. Am. J. Manage. Sci. 29, 437–456.

Davison, A. C., and Hinkley, D. V. (1997). Bootstrap Methods and Their Applications. Cambridge University Press, Cambridge.

Diaconis, P., and Efron, B. (1983). Computer-intensive methods in statistics. Sci. Am. 248, 116–130.

Diaconis, P., and Holmes, S. (1994). Gray codes for randomization procedures. Stat. Comput. 4, 287–302.

Efron, B. (1979). Bootstrap methods: Another look at the jackknife. Ann. Stat. 7, 1–26.

Efron, B. (1982). The Jackknife, the Bootstrap and Other Resampling Plans. SIAM, Philadelphia.

Efron, B. (1983). Estimating the error rate of a prediction rule: Improvements on cross-validation. J. Am. Stat. Assoc. 78, 316–331.

Efron, B., and Gong, G. (1983). A leisurely look at the bootstrap, the jackknife and cross-validation. Am. Stat. 37, 36–48.

Efron, B., and Tibshirani, R. (1986). Bootstrap methods for standard errors, confidence intervals and other measures of statistical accuracy. Stat. Sci. 1, 54–77.

Efron, B., and Tibshirani, R. (1997). Improvements on cross-validation: The 632+ rule. J. Am. Stat. Assoc. 92, 548–560.

Fisher, R. A. (1935). The Design of Experiments. Hafner, New York.

Fukuchi, J. I. (1994). Bootstrapping extremes of random variables. PhD dissertation, Iowa State University, Ames.

Gine, E., and Zinn, J. (1989). Necessary conditions for bootstrap of the mean. Ann. Stat. 17, 684–691.

Gong, G. (1986). Cross-validation, the jackknife, and the bootstrap: Excess error in forward logistic regression. J. Am. Statist. Assoc. 81, 108–113.

Good, P. I. (1994). Permutation Tests. Springer-Verlag, New York.

Gray, H. L., and Schucany, W. R. (1972). The Generalized Jackknife Statistic. Marcel Dekker, New York.

Hall, P. (1986). On the bootstrap and confidence intervals. Ann. Stat. 14, 1431–1452.

Hall, P. (1988). Theoretical comparison of bootstrap confidence intervals (with discussion). Ann. Stat. 16, 927–985.

Hall, P. (1992). The Bootstrap and Edgeworth Expansion, Springer-Verlag, New York.

Hall, P., Hardle, W., and Simar, L. (1993). On the consistency of bootstrap distribution estimators. Comput. Stat. Data Anal. 16, 11–18.

Hall, P., and Martin, M. A. (1988). On bootstrap resampling and iteration. Biometrika 75, 661–671.

Hartigan, J. A. (1969). Using subsample values as typical values. J. Am. Stat. Assoc. 64, 1303–1317.

Hartigan, J. A. (1971). Error analysis by replaced samples. J. R. Stat. Soc. B 33, 98–113.

Hartigan, J. A. (1975). Necessary and sufficient conditions for the asymptotic joint normality of a statistic and its subsample values. Ann. Stat. 3, 573–580.

Jain, A. K., Dubes, R. C., and Chen, C. (1987). Bootstrap techniques for error estimation. IEEE Trans. Pattern Anal. Mach. Intell. PAMI-9, 628–633.

Knight, K. (1989). On the bootstrap of the sample mean in the infinite variance case. Ann. Stat. 17, 1168–1175.

Lachenbruch, P. A., and Mickey, M. R. (1968). Estimation of error rates in discriminant analysis. Technometrics 10, 1–11.

Lahiri, S. N. (2003). Resampling Methods for Dependent Data. Springer-Verlag, New York.

Martin, M. A. (1990). On the bootstrap iteration for coverage correction in confidence intervals. J. Am. Stat. Assoc. 85, 1105–1118.

McCarthy, P. J. (1969). Pseudo-replication: Half-samples. Int. Stat. Rev. 37, 239–263.

McLachlan, G. J. (1980). The efficiency of Efron's bootstrap approach applied to error rate estimation in discriminant analysis. J. Stat. Comput. Simul. 11, 273–279.

Miller, R. G., Jr. (1974). The jackknife – A review. Biometrika, 61, 1–17.

Navidi, W. (1989). Edgeworth expansions for bootstrapping in regression models. Ann. Stat. 17, 1472–1478.

Pitman, E. J. G. (1937). Significance tests which may be applied to samples from any population. J. R. Stat. Soc. Suppl. 4, 119–130 and 225–232 (parts I and II).

Pitman, E. J. G. (1938). Significance tests which may be applied to samples from any population. Part III. The analysis of variance test. Biometrika 29, 322–335.

Politis, D. N., and Romano, J. P. (1994). Large sample confidence intervals based on subsamples under minimal assumptions. Ann. Stat. 22, 2031–2050.

Politis, D. N., Romano, J. P., and Wolf, M. (1999). Subsampling. Springer-Verlag, New York.

Quenouille, M. H. (1949). Approximate tests of correlation in time series. J. R. Stat. Soc. B 11, 18–84.

Shao, J., Kubler, J., and Pigeot, I. (2000). Consistency of the bootstrap procedure in individual bioequivalence. Biometrika 87, 573–585.

Simon, J. L. (1969). Basic Research Methods in Social Science. Random House, New York.

Simon, J. L., and Bruce, P. (1991). Resampling: A tool for Everyday Statistical Work. Chance 4, 22–32.

Simon, J. L., and Bruce, P. (1995). The new biostatistics of resampling. M. D. Comput. 12, 115–121.

Singh, K. (1981). On the asymptotic accuracy of Efron's bootstrap. Ann. Stat. 9, 1187–1195.

Snapinn, S. M., and Knoke, J. D. (1984). Classification error rate estimators evaluated by unconditional mean square error. Technometrics 26, 371–378.

Snapinn, S. M., and Knoke, J. D. (1985a). An evaluation of smoothed classification error rate estimators. Technometrics 27, 199–206.

Snapinn, S. M., and Knoke, J. D. (1985b). Improved classification error rate estimation. Bootstrap or smooth? Unpublished report.

Snapinn, S. M., and Knoke, J. D. (1988). Bootstrap and smoothed classification error rate estimates. Commun. Stat. Simul. Comput. 17, 1135–1153.

Tukey, J. W. (1958). Bias and confidence in not quite large samples (abstract). Ann. Math. Stat. 29, 614.

2

ESTIMATION

This chapter covers point estimation. Section 2.1 deals with ways the bootstrap can be used to estimate the bias of an estimate and " improve" it by a bias adjustment. Historically, the bootstrap was looked at to estimate the standard error of an estimate and later on for bias adjustment, while its relative the jackknife was first used for bias adjustment and then later it found a place in estimating standard errors. Section 2.1 covers bias adjustment in general, and then the application to error rate estimation in discriminant analysis is presented. Sections 2.2 and 2.3 cover measures of location and spread, respectively, as parameters to estimate by bootstrapping. Then, we look at linear regression parameters in Section 2.4 and nonlinear regression parameters in Section 2.5. Section 2.7 is the section on historical notes, and it is followed by exercises (Section 2.8) and references.

2.1 ESTIMATING BIAS

2.1.1 Bootstrap Adjustment

Let us denote the expectation of a random variable X by $E(X)$. Let $\hat{\theta}$ be an estimate of a parameter θ. Then, consider the quantity $\hat{\theta} - \theta$ to be the random variable X. The bias for the estimator $\hat{\theta}$ of θ is $b = E(X) = E(\hat{\theta} - \theta)$.

An Introduction to Bootstrap Methods with Applications to R, First Edition. Michael R. Chernick, Robert A. LaBudde.
© 2011 John Wiley & Sons, Inc. Published 2011 by John Wiley & Sons, Inc.

EXAMPLE 2.1

As an example, let us consider for univariate Gaussian random variables the maximum likelihood estimator of the variance σ^2 based on a sample of size n. The estimator

$$S_n^2 = \sum_{i=1}^{n} (X_i - X_b)^2 / n,$$

where $X_b = \sum_{i=1}^{n} X_i / n$, is the maximum likelihood estimate of σ^2.

It has a bias of $b = -\sigma^2/n$. The bootstrap estimate for the bias is b, B^* is given by $B^* = E(\theta^* - \hat{\theta})$ where $\hat{\theta}$ is S_n^2 and $\theta^* = \sum_{i=1}^{n} (X_i^* - X_b)^2 / n$. The Monte Carlo approximation to B^* is

$$B_{\text{MONTE}} = \sum B_j^* / N,$$

where B_j^* is the estimate of bias for the jth bootstrap sample for $1 \leq j \leq N$ where N is the number of bootstrap samples generated and $B_j^* = \theta_j^* - \hat{\theta}$:

$$\theta_j^* = \sum_{i=1}^{n} (X_{ij}^* - X_b)^2 / n.$$

X_{ij}^* is the ith observation in the jth bootstrap sample. Using R,

```
> set.seed(5^13) #set random seed so example reproducible
 > n<- 25 #small sample
 > x<- rnorm(n) #random sample of unit normal variate
 > x
 [1] -1.49183947 0.77485410 1.07777893 -0.18091864
1.10604087 -0.49095184 -0.83397332
 [8] -0.24173291 0.03815104 -0.11851148 -0.02272159
-0.60512628 1.93797388 -1.36609140
 [15] 1.97550278 -1.18866470 0.15011701 0.30872246
0.22374324 1.68060365 -0.71136845
 [22] -0.33415203 0.54372444 -0.55622046 0.16226317
 > varx<- var(x)*(n-1)/n #sample variance, uncorrected
 > c(varx, varx - 1.0, -1/n) #sample variance and
bias relative to true value of 1.0 and expected value of
bias
 [1] 0.8665871 -0.1334129 -0.0400000
 >
 > N<- 5000 #number of bootstrap resamples
 > bvarx<- NULL #initialize resample variances vector
 > for (i in 1:N) { #for each resample
```

```
 + xstar<- sample(x, n, replace=TRUE) #generate resample
of size n from data
 + bvarx[i]<- var(xstar)*(n-1)/n #resample variance,
uncorrected
 + }
 > thetastar<- mean(bvarx) #estimate of variance
 > c(thetastar, thetastar - varx) #resample variance
estimate and bias estimate
  [1]  0.82884859 -0.03773855
```

Generally, the purpose of estimating bias is to improve a biased estimator by subtracting an estimator of bias from it. This will reduce the bias while increasing the variance of the estimator. Bias adjustment works when the reduction in the square of the bias is greater than the increase in the variance. In that case, the mean square error (a measure of accuracy for an estimator) is reduced. However, it is possible that the variance increases too much and the bias-adjusted estimate is less accurate than the original. So, bias correction for an estimate must be done with care. The jackknife was originally created to reduce bias.

The next section will cover error rate estimation for linear discriminant functions. An estimator called resubstitution has a large bias and small variance when the sample size for constructing the classifier (i.e., classification rule) is small. It is therefore an ideal estimator for bias correction. The resubstitution estimator of error rate is obtained by fitting the classification rule using all the data. Then all the data used in the fit are reused to classify the data. The resubstitution error rates are simply the estimates obtained by counting how many classification errors are made for samples from each class divided by the number of samples in the class.

2.1.2 Error Rate Estimation in Discriminant Analysis

2.1.2.1 Introduction to Classification.
Before introducing error rate estimates and their bootstrap bias adjustment, we will provide a brief introduction to the classification problem with two classes. We then cover some of the traditional methods for estimating error rates. Then we will introduce some bootstrap methods for adjusting the bias of the resubstitution estimator and show through the results of simulation studies which methods are the most accurate.

For simplicity, we concentrate on the two-class problem. In this case there are two classes that you need to discriminate between based on a set of predictor variables and a training set (i.e., subset of data used to build the classifier). In the training set, the classes are given and that information is used to construct a classifier.

An example that Chernick worked with a great deal in the 1980s as part of President Reagan's strategic defense initiative (SDI, aka Star Wars) is the problem of telling which objects are balloon decoys and which are nuclear warheads in a reentry vehicle (RV). The balloons are discriminated from the RVs by infrared sensors on the defense satellites during the missiles' late midcourse phase when the decoys and RV are ejected from the missile. Because the RV is hotter than the balloon decoy, it should have a different spectral signature compared with the decoys. Several predictor variables from the object's signature are then used to construct the discrimination algorithm. In the literature, these variables are often referred to as features. The distribution of the features for RVs and the distribution of the features for decoys are called the class conditional densities.

Now, it is rare that a discriminant algorithm is perfect. So the rule makes some errors. It can classify a decoy as an RV or an RV as a decoy. So how do we use these error rates to find a " best" algorithm? Often a straight average of the two error rates is used as the rate to be minimized. But that treats both error rates equally.

We call this average error rate the loss function. But would you want to treat these two error rates with equal importance? If you classify a decoy as an RV, you will shoot at a decoy and waste ammunition, which can be costly. But if you misclassify an RV as a decoy, you will let a warhead reach land and thus allow a nuclear explosion. So, misclassifying an RV seems to have more serious consequence than misclassifying a decoy. So, the loss function is a weighted average of the two error rates putting more weight against committing an error classifying an RV as a decoy than classifying a decoy as an RV.

It will help with the understanding of the error rate estimator's properties if we introduce a bit of the theory for the two-class problem. For a more detailed treatment of the topic, the reader may want to consult Duda and Hart (1973), Srivastava and Carter (1983, pp. 231–253), Fukunaga (1990), or McLachlan (1992).

For simplicity, before we talk about fitting distributions based on a training set, let us consider what we would do if the distributions were known and overlap. These distributions that we are interested in are the distributions we would get when we know the classes. So we call them the class conditional densities.

Now we are considering the problem of classifying the targets versus decoys and we obtain a new feature vector but we do not know which class it comes from. What should we do? We have the problem framed as a multivariate hypothesis testing problem. We have two possible choices. Classify the object as a decoy or as a target (RV). There are two types of errors that could be made depending on the classification. If (1) we classify it as a decoy and it is actually a target or (2) we classify it as a target when it is really a decoy, we make an error.

Typically, we might follow the decision theory approach and assign costs to making the decision. The problem then becomes one of minimizing the expected unconditional cost. This is done by application of Bayes' theorem. For the target versus decoy problem, we know there are more severe consequences for classifying a target as a decoy than for classifying a decoy as a target. So realistically, the equal cost rule would not be appropriate.

However, if we do not have a problem with equal costs, the rule is simple and intuitive. We classify the object as a decoy if the a posteriori probability that the object is a decoy given the observed feature vector x is higher than the a posteriori probability that the object is a target. If this is not the case, we classify the object as a target.

Bayes' theorem provides the method for determining this. In the special case when before we collect the data we have no prior opinion favoring one class over the other, the Bayes' rule is equivalent to the frequentist's test known as the likelihood ratio test. The likelihood ratio test looks at the ratio of the class conditional densities and picks target or decoy depending on whether or not this ratio is greater than 1.

In the case where RVs are the targets and the balloon replicas are the decoys, the enemy will typically place many balloons for each RV. Sometimes, through information from intelligence agencies like the CIA, we might know that nine balloons can be carried in each missile along with the RV. Then our a priori probability of a decoy would be 0.9 and the a priori probability of an RV would be only 0.1. This is our belief before we get the infrared wavelength signature.

2.1.2.2 Bayes Estimators for Classification Error Rates.

Let us define Bayes' rule for general prior probabilities and general costs and see how it reduces to the likelihood ratio test when the prior probabilities are both 0.5 and the costs are equal. Define $P_D(x)$ and $P_T(x)$, the class conditional density for decoys and targets, respectively, when the observed feature vector is x. Let P_1 be the a priori probability for a target and $P_2 = 1 - P_1$ is the a priori probability for a decoy. This is the case because there are only these two possible classes for the object. Also, let C_1 be the cost of classifying a decoy as a target and C_2 the cost of classifying a target as a decoy. C_1 is not necessarily equal to C_2.

Now, we also define $P(D|x)$ and $P(T|x)$ as the a posteriori probability for decoys and targets, respectively, when x is the observed feature vector. Similarly, $P(D|x) = 1 - P(T|x)$. Now Bayes' theorem simply states that $P(T|x) = P(x|T)P_1/(P(x|T)P_1 + P(x|D)P_2)$.

Bayes' rule, which minimizes the expected cost for an error, is given as follows: Classify the object as a decoy if $P(x|D)/P(x|T) > K$ and as a target otherwise where $K = (C_2P_1/C_1P_2)$. Duda and Hart (1973, pp. 10–15) provided the derivation of this.

The reader should observe that we made no parametric assumptions about the form of the class conditional densities. The form of the densities does not affect the definition of the rule but does affect the shape of the decision boundary as we shall see shortly for multivariate normal assumptions. Also, notice that if $C_1 = C_2$, $K = P_1/P_2$, and if in addition $P_2 = P_1$, then $K = 1$. So if $K = 1$, the rule classifies the object as a decoy if $P(x|D)/P(x|T) > 1$. This is precisely the likelihood ratio test. From Duda and Hart (1973, p. 31) or Chernick (2007, p. 30), the decision boundaries are illustrated when the two densities are multivariate normal.

It turns out that for densities with different mean vectors but identical covariance matrices, the decision boundary is linear. When the covariance matrices are different, the decision boundary is quadratic. This becomes important because it justified the use of linear or quadratic discriminant functions (i.e., classification rule or classifier) when the multivariate normal assumptions seem to be justified. Classification rule and classifier and discriminant rule are all synonymous terms.

In the practical case where the densities are unknown, the densities can be estimated from a training set using kernel density estimates (nonparametric setting), or the normal densities are estimated by using the estimates of the mean vector, the variances, and the covariances for each density (called plug-in rules because estimates are used in place of the parameters).

Although these plug-in rules are not optimal, they are sensible ways to estimate the decision boundary and construct the classification rule. In kernel discrimination, we call this estimator the estimated decision rule. Hand's books (1981, 1982) are good sources to learn more about the kernel methods.

For known class conditional densities, Bayes' rule can be applied and as we have seen it has theoretical optimality properties. The error rates can be calculated by integrating the class conditional density for class 1 in the acceptance region for class 2 and vice versa. When parametric class conditional densities are assumed, their plug-in density estimates (e.g., multivariate normal densities) are integrated over the appropriate rejection regions to estimate the error rates. Hills (1966) showed that these plug-in estimates are optimistically biased.

2.1.2.3 Earlier Methods in the Nonparametric Setting. In the nonparametric setting, the analogue to the plug-in estimators is resubstitution. We take the discriminant rule that we fit based on all the available data and apply it to count how many of the cases are misclassified. Note that resubstitution estimate uses the same data that was used to arrive at forming the rule, so resubstitution is likewise optimistically biased and the bias can be large when the training sample size is small.

So, how might we obtain an unbiased estimate? One simple way is called the holdout method. In doing this, you randomly hold out k observations (usually called the "test data"). You fit the discriminant rule to the remaining $n - k$ observations and only evaluate the errors on the k observations that were held out. Since the randomization makes the data in the holdout set independent of the data in the set used to fit the model, the estimates of error rates will be unbiased. But this approach brings out a dilemma. It is wasteful (statistically inefficient) to hold out training data that could improve the estimated rule (the more data used to fit, the better is the fitted rule). But on the other hand, the holdout method creates an unbiased estimate.

Lachenbruch (1967) came up with a solution that was expanded on by Lachenbruch and Mickey (1968) that was nearly fully efficient and also produced an estimator that was very nearly unbiased. What was the clever trick? They would hold out only one observation and fit the model to the remaining $n - 1$ observations. This uses practically all of the data for fitting, and since they only test the one observation left out, this procedure would be unbiased. Remember, if you reuse data that fit the classification rule in the evaluation, you will get an optimistically biased estimate of the error rates.

But a holdout method that tested only one observation would have too much variability in the error rate, which on any occasion would be estimated as either 0 or 1. So the clever trick is to repeat the procedure with a second observation left out and evaluate the errors on that one. Continue to do this until each observation has been left out once.

It seems intuitive that the estimate that averages all the errors made would be nearly unbiased and efficient because all the data are used to obtain the estimates, and the error counting is based on a discriminant function that did not include the observation being left out. This procedure is referred to as "leave-one-out" or cross-validation and became very popular because of its ability to produce nearly unbiased estimates for very general families of classification rules. The procedure is not perfectly unbiased because n different discriminant rules are used in the classification of error rates rather than just the one that would be used based on all n training vectors.

What was not appreciated until Glick (1978) was the large variance of the leave-one-out estimates. Glick introduced "smoothed" estimators to reduce this variance. This was successfully improved on in the papers by Snapinn and Knoke (1984, 1985a). Other variations include leaving blocks of a small number k of observations each time instead of one ("k-fold" cross-validation). Typically, a random 5–10% of the total data set is left out at each step.

2.1.2.4 Bootstrap Methods and Cross-Validation Compared. These estimates of Glick and Snapinn and Knoke are not bootstrap estimates, but they are superior to leave-one-out. One year earlier, Efron (1983) showed that leave-one-out (which he called cross-validation) could be greatly improved upon, with respect to mean square error, when the training set sample size is small, by bootstrap estimates, which start by taking the resubstitution estimate and adjust for its bias.

Efron's results showed that a particular estimator referred to as the 632 estimator (because it weighs one estimate, e_0, by $0.632 \approx 1 - e^{-1}$ and a second, resubstitution, by $0.368 \approx e^{-1}$) was most often the best. His simulations were for two-class problems and linear discriminant functions only using low-dimensional multivariate normal distributions with common covariance matrices. Chernick et al. (1985, 1986, 1988a, b) extended the results to more situations and to non-normal distributions and showed that the superiority of the 632 estimate goes away when the class conditional densities are heavy tailed.

The e_0 estimate is a bootstrap estimate recommended by Chatterjee and Chatterjee (1983) and later named e_0 in Efron (1983). The estimate is obtained by first generating bootstrap samples. For each bootstrap sample, a classification rule is constructed. The error rates for each class are obtained by taking the original data and counting the proportion in each class that is misclassified. This is repeated for each bootstrap sample, and the e_0 estimate is obtained by averaging the estimates over the collection of bootstrap samples. Efron (1983) described a different bootstrap estimate that he calls the bootstrap. It is the estimate we label as BOOT in the simulations. BOOT is more complicated to define. For the curious reader, we refer to Chernick (2007, p. 33) for a detailed description.

Efron and Tibshirani (1997) improved on the 632 estimator with a variant that they called "632+." Jain et al. (1987) extended Efron's results to quadratic discriminant functions simulating multivariate normal class conditional densities with unequal covariance matrices. We present here selected tables from Chernick et al. (1988b). See Chernick (2007) for a more detailed account of these simulation studies.

Table 2.1 shows the results for the parameter $M = 1.3$ and 3.0 (extremely heavy-and-light-tailed Pearson VII distributions). The Pearson VII family is a family of elliptically contoured multivariate densities and thus is similar to the multivariate normal distribution when the high-order moments exist (at least the first four). Chernick et al. (1988b) used it not only because of its similarity to the normal distribution but also because the tail behavior of the distribution is controlled simply by the parameter M. M is known and is varied in the simulations for the purpose of seeing how the shape

TABLE 2.1. Summary of Estimators Using Root Mean Square Error (Shows the Number of Simulations for Which the Estimator Attained One of the Top Three Ranks among the Set of Estimators)

Rank	632	MC	e_0	Boot	CONV	U	APP	Total
				$M = 1.3$				
First	0	0	2	0	10	0	0	12
Second	3	0	0	9	0	0	0	12
Third	0	9	0	1	2	0	0	12
Total	3	9	2	10	12	0	0	36
				$M = 3.0$				
First	21	0	6	0	0	0	3	30
Second	9	3	5	3	2	2	6	30
Third	0	8	1	8	11	1	1	30
Total	30	11	12	11	13	3	10	90

Source: Chernick et al. (1988b).

of the population distribution affects the accuracy of the competing estimates. The probability density function is given by

$$f(x) = \Gamma(M)|\Sigma|^{1/2} \Big/ \Big\{ \Gamma[M - (p/2)] \pi^{p/2} \Big[1 + (x - \mu)^T (x - \mu) \Big]^M \Big\},$$

where μ is a location vector, Σ is a scaling matrix, M is a parameter that affects the dependence among components of the vector x and the tail behavior, p is the dimension of x (in our table $p = 2$), and Γ is the gamma function. The symbol $\|$ denotes the determinant of the matrix.

The estimators are labeled 632, MC, e_0, BOOT, CONV, U, and APP. U is the leave-one-out (cross-validation) estimator and APP is resubstitution (or the apparent error rate). The others are all variants of the bootstrap with BOOT being Efron's original proposed bootstrap, 632 and e_0 as previously described, and CONV and MC are two variants proposed by Chernick, Murthy, and Nealy in their series of papers. See Chernick (2007) for the formal definitions of each.

The results show that for $M = 3.0$, 632 is the best in over two-thirds of the cases and second in the remaining nine. The others finish second or third almost equally with the exception of U, which is only in the top 3 three times. This is very similar to the results for normal distributions. Results depend somewhat on how well separated the distributions are, which is determined by the values of μ since for each distribution, μ represents the center of the Pearson VII distribution (i.e., the center of the elliptic contours). As long as we keep Σ, p, and M the same for each class and separate the distributions by moving μ_1 for class 1 and μ_2 for class 2 further apart, the error rates will decrease. In the case of $M = 1.3$, a very heavy-tailed case where no moments exist, the CONV estimator appears to be the best. The important point is that 632 is no longer a strong competitor as the opposite biases of APP and e_0 disappear.

More recently, Fu et al. (2005) provided a different approach to the problem. They took bootstrap samples to act as the data and applied cross-validation to the bootstrap samples. Based on simulations involving small sample sizes (16 or less), they found that this approach compares favorably with the bootstrap bias adjustment approaches and may be better than 632+ in some problems. Van Sanden et al (2010) measured the performance of biomarkers used to predict binary outcomes. They used bootstrap cross-validation (BCV) to estimate these error rates without the need for bias adjustment.

2.1.2.5 BCV. Because of the importance of the results of Fu et al. (2005) and the application of it by Van Sanden et al. (2011) in the area of evaluating genetic bio-markers in drug development, we shall discuss those papers in some detail. The idea of Fu et al. (2005) was to combine the ideas of bootstrap with the nice properties of cross-validation as first expounded by Stone (1974) and Geisser (1975). They were motivated by the need for accurate small sample size estimates of classification error rates in microarray studies. The work of Efron (1983) and Efron and Tibshirani (1997), which we have previously discussed, encouraged them to employ bootstrapping because of previous success in small sample sizes as we have seen in these and various other simulation studies. Braga-Neto and Dougherty (2004) looked at cross-validation in the context of training based on a small number of genes that investigators find interesting. On the other hand, Ambroise and McLachlan (2002) applied a 10-fold cross-validation, and a bootstrap 632+, to accurate microarray data where the number of genes is very large (over 1000) and the sample size is still small. Given these research results, Fu, Carroll, and Wang had the idea of looking at whether or not bootstrap resampling could be used to improve cross-validation. Comparisons are made using mean square relative error (MSRE) as the criterion.

Using the notation of Fu et al. (2005), we take the sample to be

$$S = \{(x_1, y_1), (x_2, y_2), \ldots, (x_n, y_n)\},$$

where x_i is the observed feature value (or feature vector) and y_i is the class indicator for x_i for $i = 1, 2, \ldots, n$. From these data, they generate bootstrap samples of size n with replacement. Let B be the number of bootstrap replications, and for each replication, we denote the bootstrap sample by

$$S_b^* = \{(x_1^*, y_1^*), (x_2^*, y_2^*), \ldots, (x_n^*, y_n^*)\},$$

for $b = 1, 2, \ldots, n$. For each bootstrap sample, a specific predetermined rule (e.g., linear discriminant analysis [LDA]), a cross-validation estimate is used to evaluate the prediction rule. In each case, an estimate of the error rate is computed as r_b and the BCV estimate of the error rate is $r_{\mathrm{BCV}} = B^{-1} \sum_{b=1}^{B} r_b$, the average across bootstrap samples. The authors used a slight modification of the bootstrap by restricting the sample to have at least three distinct observations in each class. This is because two or fewer in a training set makes it difficult to construct a good classifier that would be representative of what you would see practically in a sample of size n. In Section 2.6.2, we will discuss Breiman's bagging methodology. Since BCV is an average of cross-validated estimates of error rates, a form of a bootstrap aggregation, it can be viewed as a bagging procedure (Breiman, 1996). Although Breiman did bagging in the context of averaging prediction or classification algorithms to improve the prediction, we are doing essentially the same thing here, except on the algorithm to estimate the accuracy of a single prediction algorithm. The authors supplied a heuristic rationale as to why BCV is better than cross-validation (CV) in small samples. Basically, the bootstrap samples tend to include duplicates. That tends to reduce the variability of the cross-validated estimates (and averaging helps even further). The CV estimates are practically unbiased, and since we are averaging over bootstrap samples, the BCV estimate will not have any increase in bias. The argument regarding 632 and 632+ is that they put a large weight on the e_0 estimate causing some bias that is larger in small samples than what we have for BCV.

However, the simulation results are more persuasive. After all, the proof of the pudding is in the eating. The authors provided us with four simulation scenarios.

Case 1: two-class problem with equal sample sizes and class 1 observations taken from a normal $(0, 1)$ distribution and class 2 observations from normal $(\Delta, 1)$ with $\Delta > 0$ and sample sizes of $n/2$ in each group with $n = 20, 30, 50,$ and 100. In each case, $B = 200$. The classification algorithm was a quadratic discriminant analysis (QDA).

The mean relative error denoted as $R(S) = \{r - r(S)\}/r(S)$ where r is the true conditional error rate given the rule and the class conditional distributions, and $r(S)$ is the estimate of that error rate for a given estimator and sample S. The simulations were repeated 1000 times, and the performance measure is the average of the squares of the values of $R(S_i)$ for samples S_i with $i = 1, 2, \ldots, 1000$. For $\Delta = 1$ and $\Delta = 3$. $\text{MSRE} = 1000^{-1} \sum R(S_i)^2$ is compared with BCV, CV, e_0, and 632 for each sample size. See Table 2.2 extracted from table 1 of Fu et al. (2005).

We note from the table that BCV is best (i.e., has the smallest MSRE) at each sample size and 632 beats CV and e_0 as expected. However, it is not as good as BCV, although it gets closer as the sample size increases.

For $\Delta = 3$, the separation between the groups is greater than for $\Delta = 1$, and hence, the classifier has a lower error rate but the mean relative error increases. The interpretation of this is that as the true error rate gets smaller, the various conditional error rate

TABLE 2.2. Square Root of MSRE for QDA Normal (0, 1) versus Normal (Δ, 1) (Case 1)

Δ	N	BCV	CV	e_0	632
1	20	0.2828	0.4171	0.3926	0.3314
	30	0.2452	0.2917	0.3265	0.2805
	50	0.2098	0.2293	0.2557	0.2283
	100	0.1480	0.1559	0.1612	0.1543
3	20	0.7290	0.8387	0.9557	0.8233
	30	0.6319	0.7281	0.7495	0.6867
	50	0.4979	0.5552	0.5372	0.5177
	100	0.3719	0.4061	0.3878	0.3847

TABLE 2.3. Square Root of MSRE for Five Nearest-Neighbor Classifier Chi-Square 1 Degree of Freedom (df) versus Chi-Square 3 df and Chi-Square 5 df (Equal Sample Sizes of $n/2$) (Case 3)

df for Class 2	n	BCV	CV	e_0	632
3	16	0.3126	0.5235	0.4672	0.3520
	20	0.2718	0.4390	0.4153	0.3124
	30	0.2375	0.3497	0.3165	0.2480
	50	0.2093	0.2739	0.2415	0.1921
5	16	0.4869	0.6515	0.3234	0.5265
	20	0.4032	0.5360	0.2662	0.4559
	30	0.3335	0.4815	0.2230	0.3782
	50	0.2828	0.3778	0.1697	0.3082

estimates get relatively larger. This means that it is more difficult to accurately estimate the smaller error rates in a relative sense.

Case 2 differs from case 1 only in that unequal sample sizes are used and five-fold cross-validation is added for comparison. This shows the very poor performance of fivefold cross-validation and provides justification for not including it in subsequent simulations. Again, BCV is best with 632 second when $n = 20$, but not for 30 or 50 where e_0 is slightly better.

In case 3, a skewed distribution was considered comparing a chi-square distribution with 1 degree of freedom for class 1 to a chi-square with 3 and with 5 for class 2. Results in Table 2.3 were extracted from table 3 in Fu et al. (2005).

Although we are looking at skewed distributions and a nearest neighbor algorithm instead of a QDA, the pattern is the same for examples 3 and 4 (Table 2.4 and 2.5) with BCV being the best for all chosen sample sizes and 632 next. Again, the difference between 632 and BCV gets smaller as the sample size increases.

In case 4, multidimensional normal distributions are taken for the class conditional densities for dimension $q = 5$ and 10. As the dimension q increases, the data tend to spread out away from the center so the algorithm chosen was to decide based on only the three nearest neighbors rather than five. Results extracted from table 4 of Fu et al. (2005) are given in Table 2.4.

Once again the pattern remains the same even though the algorithm changed and multivariate (though uncorrelated) normal distributions were used.

For the microarray application, data were taken from van't Veer et al. (2002) and van de Vijver et al. (2002) where gene profiling was done on 295 patients for breast cancer based on 70 genes. By reducing the set of genes to the five that appear to be most closely related to the prognosis, the class conditional densities could be closely approximated to give a class conditional error rate that they could use as a standard (being close to the truth). Two classification schemes were used (K nearest neighbor [KNN], with $K = 3$ and with $K = 5$).

Relative errors were calculated relative to the estimates based on the class conditional densities. The choice of 5 genes out of 70 using the best correlation is known to be biased (e.g., Ambroise and McLachlan, 2002). But the point of the comparison is to

TABLE 2.4. Square Root of MSRE for Three Nearest-Neighbor Classifier Normal Mean = 0, 5 dim. and 10 dim. versus Normal Mean = Δ5 dim. and 10 dim. (Equal Sample Sizes of $n/2$) (Case 4)

Dim. and separation (q, Δ), respectively for class 2	n	BCV	CV	e_0	632
(5, 1)	16	0.3481	0.5918	0.5926	0.3932
	20	0.2934	0.5198	0.5424	0.3373
	30	0.2472	0.3805	0.4277	0.2640
(10, 0.8)	16	0.3291	0.5892	0.6051	0.3835
	20	0.2857	0.5003	0.5948	0.3410
	30	0.2540	0.4291	0.5243	0.3041

TABLE 2.5. Square Root of MSRE for Three and Five Nearest-Neighbor Classifiers with Real Microarray Data (Equal Sample Sizes of $n/2$)

Number of nearest neighbors	n	BCV	CV	e_0	632
3	20	0.3183	0.4734	0.4528	0.3324
	30	0.3003	0.3835	0.3568	0.2872
5	20	0.3208	0.4702	0.4761	0.3564
	30	0.2879	0.3894	0.4071	0.3156

Results are similar except that CV does better than e_0 when $K = 5$.

see how the error rate estimation algorithm works in a small enough dimensional space that the class conditional error rates could be well approximated as a standard for the mean relative error rates. The objective was not necessarily to pick the best set of genes. That would be the objective when the error rate estimation methodology is fixed and used to help select the best genes.

The comparison of the estimates of error rates for the microarray data is extracted from table 5 of Fu et al. (2005) and given in Table 2.5.

Now we turn to the article by Van Sanden et al. (2011). They are interested in determining if certain types of genes can serve as biomarkers for a binary outcome variable. The type of genes being considered are those that are related to treatment, those related to response, and those that are related to both treatment and response. Their idea is to extend the joint model that was introduced by Lin et al. (2010) designed for continuous response, to handle a binary outcome variable.

The use of gene expression values to predict specific responses is now well-documented in the literature. The results are based on microarray experiments that provide a tool for the identification of genetic biomarkers. In their paper, Van Sanden et al. considered experiments where the treatment was administered to the subjects before response and gene expression data are collected. Their intent is to look at and possibly discover relationships between the gene expression measurements, the treatment, and the outcome of primary interest. Their study has many objectives. In order to better understand the biological processes that take place as a result of the treatment, the research scientist wants to identify genes that are affected by the drug. Another possibility is to find genes that are related to responses due to phenotype. Examples are toxicity and efficacy.

One goal is to evaluate whether or not a particular gene can serve as a biomarker for the primary outcome variable in a clinical trial. This would mean that the relationship is strong enough to predict phenotypic response. Such genes would represent potential therapeutic biomarkers. The authors analyzed data from two preclinical studies. One is a behavioral study and the other is a toxicology study.

In the behavioral study, an experiment was performed on rats concerning compulsive checking disorder (Szechtman et al., 2001). The rat's home base is defined as the location where the animal spends the longest total amount of time. The data were counts on how often the rat returned to his home base from out in the open field. These rats were classified as having the disorder if the treatment causes an increase in the number of returns to home after treatment. This is a binary outcome since the count either increases or it does not. The median of the distribution of home base visits determines who the responders are since an increase in return visits classifies them as responders and any other result does not.

In the toxicology study, kidney vacuolation sometimes occurs when Sprague–Dawley rats are treated with certain compounds. The goal of the study was to find changes in gene expression following 28 days of treatment. One hundred rats were randomized to three treatment groups and one control group. The authors considered two modeling approaches to the problem.

One involves testing multiple hypotheses with multiplicity adjustment. But this is not a particularly promising approach for binary outcomes and cannot be easily extended to more than two categories. They also found that the algorithm had convergence problems in the toxicity study and it is very computationally intensive. The other is a selection algorithm that ranks the genes based on what is called the BW criterion. BW stands for the ratio of between-group sum of squares to within-group sum of squares. They follow the approach of Dudoit et al. (2002) and Van Sanden et al. (2007). In both papers, diagonal linear discriminant analysis (DLDA) is used, where the authors have shown that under the conditions of most microarray problems, DLDA works better than classification trees and support vector machines. After ranking, the DLDA classification algorithm is constructed. In their approach, the authors applied the BCV method of Fu et al. (2005). Both approaches lead to identification of potential biomarkers that are related to treatment, response, or both.

2.1.3 Simple Example of Linear Discrimination and Bootstrap Error Rate Estimation

To illustrate what is done in these classification problems, we provide a simple example. Suppose that we have five bivariate normal training vectors from class 1 and five from class 2. Class 1 has a bivariate normal distribution with mean $(0, 0)^T$ and covariance matrix equal to the identity matrix, while class 2 has a bivariate normal distribution with mean $(1, 1)^T$ and a covariance matrix that is also the identity matrix. The training vectors were generated from the assumed distributions and are as follows:

TABLE 2.6. Truth Tables for the Six Bootstrap Samples

True class	Sample #1 classified as		Sample #2 classified as	
	Class 1	Class 2	Class 1	Class 2
Class 1	4	1	5	0
Class 2	0	5	0	5
True class	Sample #3 classified as		Sample #4 classified as	
	Class 1	Class 2	Class 1	Class 2
Class 1	4	0	6	0
Class 2	0	6	0	4
True class	Sample #5 classified as		Sample #6 classified as	
	Class 1	Class 2	Class 1	Class 2
Class 1	4	0	2	0
Class 2	0	6	0	8

TABLE 2.7. Resubstitution Truth Table for Original Data

True class	Original sample classified as	
	Class 1	Class 2
Class 1	5	0
Class 2	0	5

- For class 1, $(2.052, 0.339)^T$, $(1.083, -1.320)^T$, $(0.083, -1.524)^T$, $(1.278, -0.459)^T$, and $(-1.226, -0.606)^T$.
- For class 2, $(1.307, 2.268)^T$, $(-0.548, 1.741)^T$, $(2.498, 0.813)^T$, $(0.832, 1.409)^T$, and $(1.498, 2.063)^T$.

We shall generate six bootstrap samples of size 10 and calculate the standard bootstrap estimate of the error rate using bias correction and also the 632 estimate, which is a linear combination of e_0 and resubstitution (Tables 2.6 and 2.7). We shall denote by the indices 1, 2, 3, 4, and 5 for the five samples from class 1 as ordered above and 6, 7, 8, 9, and 10 for the five samples from class 2 also in the order given above.

There are really two ways to bootstrap here. One would be to stratify so that we sample five with replacement from class 1, and five with replacement from class 2. In Efron (1983), he did not stratify, and others chose to do it the same way as Efron for comparison purposes. The bootstrap without stratification would simply sample the indices from 1 to 10 randomly with replacement. This allows an imbalance where say 6 or 7 of the indices are from class 1 with only 4 or 3 from class 2, respectively. Of course, the imbalance can just as easily occur the opposite way. The choice of method does not make a big difference unless 9 or all 10 cases come from one of the classes and that is not likely to happen. In our case, the sample indices are (9,3,10,8,1,9,3,5,2,6),

TABLE 2.8. Linear Discriminant Function Coefficients for Original Data

Class number	Constant term	Variable no. 1	Variable no. 2
1	−1.49338	0.56332	−1.72557
2	−4.40375	0.57400	3.65311

TABLE 2.9. Linear Discriminant Function Coefficients for Bootstrap Samples

Class number	Constant term	Variable no. 1	Variable no. 2
	Bootstrap sample #1		
1	−1.79304	0.68513	−2.06630
2	−3.78089	1.02701	2.97943
	Bootstrap sample #2		
1	−2.42499	1.00925	−3.28463
2	−3.98211	−0.19893	4.53327
	Bootstrap sample #3		
1	−2.56595	−0.17167	−3.42961
2	−6.40871	1.34040	6.54905
	Bootstrap sample #4		
1	−1.52443	0.65570	−2.59189
2	−6.35297	−0.46918	6.00759
	Bootstrap sample #5		
1	−3.79341	0.43677	−4.53972
2	−6.39451	1.02171	6.52275
	Bootstrap sample #6		
1	−3.10983	−1.31926	−1.10969
2	−3.98019	2.14605	3.30526

(1,5,7,9,9,9,2,3,3,9), (6,4,3,9,2,8,7,6,7,5), (5,5,2,7,4,3,6,9,10,1), (2,6,4,8,8,10,3,3,7,7), and (6,7,3,6,10,10,7,8,5,9).

We notice that only bootstrap sample six was highly imbalanced with 2 from class 1 and 8 from class 2. Bootstrap sample one had 5 from class 1 and 5 from class 2. Sample two had 5 from class 1 and 5 from class 2. Sample three had 4 from class 1 and 6 from class 2. Sample four had 6 from class 1 and 4 from class 2. Sample five had 4 from class 1 and 6 from class 2. The (linear discriminant) coefficients for the original data are given in Table 2.8 and for the bootstrap samples in Table 2.9.

We now provide the SAS code using Version 9.2 that calculated these discriminant functions:

```
data classes;
input Class_no $ x1 x2;
datalines;
Class_1 2.052 0.339
Class_1 1.083 -1.320
Class_1 0.083 -1.524
Class_1 1.278 -0.459
Class_1 -1.226 -0.606
Class_21.307 2.268
Class_2-0.548 1.741
Class_22.498 0.813
Class_20.832 1.409
Class_21.498 2.063
;
run;
title 'The Original Sample Discriminant Function';
proc discrim data=classes outstat=class_stat
method=normal pool=yes
list crossvalidate;
class Class_no;
priors prop;
var x1 x2;
run;
proc print data=classes;
run;
data class_boot1;
input Class_no $ x1 x2;
datalines;
Class_1 2.052 0.339
Class_1 1.083 -1.320
Class_1 0.083 -1.524
Class_1 0.083 -1.524
Class_1 -1.226 -0.606
Class_2 1.307 2.268
Class_20.832 1.409
Class_2 2.498 0.813
Class_2 0.832 1.409
Class_2 1.498 2.063
;
```

```
run;
title 'The First Bootstrap Sample Discriminant Function';
proc discrim data=class_boot1 outstat=class_stat
method=normal pool=yes
list crossvalidate;
class Class_no;
priors prop;
var x1 x2;
run;
proc print data=class_stat;
run;
data class_boot2;
input Class_no $ x1 x2;
datalines;
Class_1 2.052 0.339
Class_11.083 -1.320
Class_1 0.083 -1.524
Class_1 0.083 -1.524
Class_1 -1.226 -0.606
Class_2 -0.548 1.741
Class_20.832 1.409
Class_2 0.832 1.409
Class_2 0.832 1.409
Class_2 0.832 1.409
;
run;
title 'The Second Bootstrap Sample Discriminant Function';
proc discrim data=class_boot2 outstat=class_stat
method=normal pool=yes
list crossvalidate;
class Class_no;
priors prop;
var x1 x2;
run;
proc print data=class_stat;
run;
data class_boot3;
input Class_no $ x1 x2;
datalines;
Class_1 1.083 -1.320
```

```
Class_1 0.083 -1.524
Class_1 1.278 -0.459
Class_1 -1.226 -0.606
Class_2 1.307 2.268
Class_2 1.307 2.268
Class_2 -0.548 1.741
Class_2 -0.548 1.741
Class_2 2.498 0.813
Class_2 0.832 1.409
;
run;
title 'The Third Bootstrap Sample Discriminant Function';
proc discrim data=class_boot3 outstat=class_stat
method=normal pool=yes
list crossvalidate;
class Class_no;
priors prop;
var x1 x2;
run;
proc print data=class_stat;
run;
data crop_boot4;
input Class_no $ x1 x2;
datalines;
Class_1 2.052 0.339
Class_1 1.083 -1.320
Class_1 0.083 -1.524
Class_1 1.278 -0.459
Class_1 -1.226 -0.606
Class_1 -1.226 -0.606
Class_2 1.307 2.268
Class_2 -0.548 1.741
Class_2 0.832 1.409
Class_2 1.498 2.063
;
run;
title 'The Fourth Bootstrap Sample Discriminant Function';
proc discrim data=class_boot4 outstat=class_stat
method=normal pool=yes
list crossvalidate;
```

```
class Class_no;
priors prop;
var x1 x2;
run;
proc print data=class_stat;
run;
data class_boot5;
input Class_no $ x1 x2;
datalines;
Class_1 1.083 -1.320
Class_1 0.083 -1.524
Class_1 0.083 -1.524
Class_1 1.278 -0.459
Class_2 1.307 2.268
Class_2 -0.548 1.741
Class_2 -0.548 1.741
Class_2 2.498 0.813
Class_2 0.832 1.409
Class_2 1.498 2.063
;
run;
title 'The Fifth Bootstrap Sample Discriminant Function';
proc discrim data=class_boot5 outstat=class_stat
method=normal pool=yes
list crossvalidate;
class Class_no;
priors prop;
var x1 x2;
run;
proc print data=class_stat;
run;
data class_boot6;
input Class_no $ x1 x2;
datalines;
Class_1 0.083 -1.524
Class_1 -1.226 -0.606
Class_2 1.307 2.268
Class_2 1.307 2.268
Class_2 -0.548 1.741
```

```
Class_2 -0.548 1.741
Class_2 2.498 -0.459
Class_2 0.832 1.409
Class_2 1.498 2.063
Class_2 1.498 2.063
;
run;
title 'The Sixth Bootstrap Sample Discriminant Function';
proc discrim data=class_boot6 outstat=class_stat
method=normal pool=yes
list crossvalidate;
class Class_no;
priors prop;
var x1 x2;
run;
proc print data=class_stat;
run;
```

A roughly equivalent R code:

```
#The Original Sample Discriminant Function
classes<- data.frame(class=c(rep(1,5),rep(2,5)),
x1=c(2.052,1.083,0.083,1.278,-1.226, 1.307,
-0.548,2.498,0.832,1.498),

x2=c(0.339,-1.320,-1.524,-0.459,-0.606,
2.268,1.741,0.813,1.409,2.063))
classes #show data
require('MASS')
ldaOrig<- lda(class ~x1 + x2, data=classes)
ldaOrig
predict(ldaOrig, classes)$class #show resubstitution
assignment
#First Bootstrap Sample Discriminant Function
classes1<- data.frame(class=c(rep(1,5),rep(2,5)),
x1=c(2.052,1.083,0.083,0.083,-1.226,
1.307,0.832,2.498,0.832,1.498),

x2=c(0.339,-1.320,-1.524,-1.524,-0.606,
2.268,1.409,0.813,1.409,2.063))
classes1 #show data
```

```
ldaBS1<- lda(class ~x1 + x2, data=classes1)
ldaBS1
predict(ldaBS1, classes)$class #show performance on
original data
#Second Bootstrap Sample Discriminant Function
classes2<- data.frame(class=c(rep(1,5),rep(2,5)),
x1=c(2.052,1.083,0.083,0.083,-1.226,
-0.548,0.832,0.832,0.832,0.832),

x2=c(0.339,-1.320,-1.524,-1.524,-0.606,
1.741,1.409,1.409,1.409,1.409))
classes2 #show data
ldaBS2<- lda(class ~1 + x2, data=classes1)
ldaBS2
predict(ldaBS2, classes)$class #show performance on
original data
#Third Bootstrap Sample Discriminant Function
classes3<- data.frame(class=c(rep(1,4),rep(2,6)),
x1=c(1.083,0.083,1.278,-1.226,1.307, 1.307,-0.548,
-0.548,2.498,0.832),

x2=c(-1.320,-1.524,-0.459,-0.606,2.268,
2.268,1.741,1.741,0.813,1.409))
classes3 #show data
ldaBS3<- lda(class ~x1 + x2, data=classes3)
ldaBS3
predict(ldaBS3, classes)$class #show performance on
original data
#Fourth Bootstrap Sample Discriminant Function
classes4<- data.frame(class=c(rep(1,6),rep(2,4)),
x1=c(2.052,1.083,0.083,1.278,-1.226,-1.226, 1.307,
-0.548,0.832,1.498),

x2=c(0.339,-1.320,-1.524,-0.459,-0.606,-0.606,
2.268,1.741,1.409,2.063))
classes4 #show data
ldaBS4<- lda(class ~x1 + x2, data=classes4)
ldaBS4
predict(ldaBS4, classes)$class #show performance on
original data
#Fifth Bootstrap Sample Discriminant Function
classes5<- data.frame(class=c(rep(1,4),rep(2,6)),
```

```
x1=c(1.083,0.083,0.083,1.278,  1.307,-0.548,
-0.548,2.498,0.832,1.498),


x2=c(-1.320,-1.524,-1.524,-0.459,
2.268,1.741,1.741,0.813,1.409,2.063))
classes5 #show data
ldaBS5<- lda(class ~x1 + x2, data=classes5)
ldaBS5
predict(ldaBS5, classes)$class #show performance on
original data
#Sixth Bootstrap Sample Discriminant Function
classes6<- data.frame(class=c(rep(1,2),rep(2,8)),
x1=c(0.083,-1.226,  1.307,1.307,-0.548,
-0.548,2.498,0.832,1.498,1.498),


x2=c(-1.524,-0.606,  2.268,2.268,1.741,1.741,
-0.459,1.409,2.063,2.063))
classes6 #show data
ldaBS6<- lda(class ~x1 + x2, data=classes6)
ldaBS6
predict(ldaBS6, classes)$class #show performance on
original data
```

2.1.4 Patch Data Example

When we make comparisons, we may be interested in a ratio of two variables instead of a difference. Suppose we have data that enable us to compute unbiased estimates for each variable. How should we estimate the ratio? One natural inclination would be to just take the ratio of the two unbiased estimates. But, unfortunately, such an estimate will be biased, except in degenerate cases.

To explain this, suppose that the random variable X produces an unbiased estimate of θ and Y an unbiased estimate of μ. So, $E(X) = \theta$ and $E(Y) = \mu$. So $E(X)/E(Y) = \theta/\mu$. But if we use X/Y to be the estimate, we are interested in $E(X/Y)$, which is usually not equal to $E(X)/E(Y) = \theta/\mu$. Now, if we go one step further and assume X and Y are statistically independent, then $E(X/Y) = E(X)E(1/Y) = \theta E(1/Y)$. Now, the reciprocal function $f(z) = 1/z$ is a convex function. So by Jensen's inequality (see e.g., Ferguson, 1967, pp. 76–78), $f(E(Y)) = 1/\mu \leq E(f(Y)) = E(1/Y)$. This shows that $E(X/Y) = \theta E(1/Y) \geq \theta/\mu$. This inequality is strict except in the degenerate case when Y is a constant. So there is a positive bias $B = E(X/Y) - \theta/\mu$. This bias can be large and when it is, it is natural to try to improve on it by a bias adjustment.

Efron and Tibshirani (1994) provided an example where bootstrapping to estimate or adjust for bias could be attempted. They call it the patch data example because the data are the amount of hormone in a patch that is placed on the patient. This was a small clinical trial to establish that a hormone patch developed at a new manufacturing site was equivalent to the patch developed at a previously approved site.

The U.S. Food and Drug Administration (FDA) has a well-defined criterion for establishing bioequivalence for such trials. The manufacturer must show that the new hormone (hormone produced at the new plant) produces at least 80% of the effect of the old hormone (hormone produced at the approved plant) over placebo. But the new hormone will be tested against the old hormone and compared with placebo only through the use of the data from a previous trial where the old hormone was tested against placebo.

Mathematically, we define the following ratio as the parameter we are interested in. Let $\theta = \{E(\text{new}) - E(\text{old})\}/\{E(\text{old}) - E(\text{placebo})\}$ where $E(\text{new})$ is the expected level of hormone in the patients' bloodstream from the hormone produced at the new plant, $E(\text{old})$ is the expected level of hormone in the patients' bloodstream from the hormone produced at the approved plant, and $E(\text{placebo})$ is the expected hormone level when a placebo patch is given. The manufacturing plant is irrelevant for placebo.

The placebo is just there to represent a baseline level of the hormone in patients with hormone deficiency when they have not been given a real treatment. The only data available to estimate this expectation come from the previous approved trial. Since an effective treatment has been established for these patients, it is not ethical to randomize them to a placebo arm. The criterion then says that we want $|\theta| > 0.20$. The reason for this is that the criterion fails only if $E(\text{new})$ is less than $E(\text{old})$ by more than 20% of $E(\text{old}) - E(\text{placebo})$. Otherwise, it is 80% or more of $E(\text{old}) - E(\text{placebo})$. Since the patch was approved, we can assume $E(\text{old}) - E(\text{placebo}) > 0$.

How did we arrive at this? The FDA criterion is that

$$|E(\text{new}) - E(\text{placebo})| \geq 0.80\{E(\text{old}) - E(\text{placebo})\},$$

but for this to happen, we know that we need $E(\text{new}) - E(\text{placebo}) > 0$ since $E(\text{old}) - E(\text{placebo}) > 0$. This means that we can remove the absolute value sign from the left-hand side of the equation, and then if we add and subtract $E(\text{old})$ from the left-hand side of the equation, we get

$$E(\text{new}) - E(\text{old}) + E(\text{old}) - E(\text{placebo}) \geq 0.80\{E(\text{old}) - E(\text{placebo})\}.$$

Now, we subtract $E(\text{old}) - E(\text{placebo})$ from both sides of the equation and get

$$E(\text{new}) - E(\text{old}) \geq -0.20\{E(\text{old}) - E(\text{placebo})\}.$$

Taking absolute values on both sides, we see that this implies

$$|E(\text{new}) - E(\text{old})| \geq 0.20|E(\text{old}) - E(\text{placebo})|$$

or

$$|E(\text{new}) - E(\text{old})|/|E(\text{old}) - E(\text{placebo})| \geq 0.20.$$

To show this, we want to reject

$$|E(\text{new}) - E(\text{old})|/|E(\text{old}) - E(\text{placebo})| < 0.20;$$

that is, $|\theta| \leq 0.20$, so our null hypothesis is

$$H_0 : |\theta| \leq 0.20.$$

Now, since θ is unknown, we need to estimate it from our data from the two trials and test the hypothesis based on the null distribution of that estimate.

How might we do this? As we saw before, we could plug in the sample mean estimates in the numerator and denominator leading to a biased estimator of θ (call this the plug-in estimator). Notice here that the situation is more complicated because in addition to the bias that occurs from taking ratios of averages, we may also have a dependence of the numerator and denominator if we use the same estimate of $E(\text{old})$ in both the numerator and denominator. This problem could be avoided if we only use data from the new trial in the numerator and only data from the old trial in the denominator.

In the example, there were eight measurements of the new patch, eight of the old and eight of the placebo. So the numerator and denominator are dependent and the bias of the plug-in estimate is needed. For their data, Efron and Tibshirani estimated θ to be -0.0713. Now, this is certainly less than 0.20 in absolute value. But it is an estimate that could have a large bias. So for hypothesis testing, we might use bootstrap confidence intervals, which will be covered in the next chapter. However, Efron and Tibshirani were content to estimate the bias by the bootstrap, and they found that the bootstrap estimate of bias was 0.0043. Since this bias is more than an order of magnitude lower than the estimate itself, it seems safe to conclude that $|\theta| \leq 0.20$. For more details, see Efron and Tibshirani (1994) or Chernick (2007).

2.2   ESTIMATING LOCATION

This section covers point estimates of location, specifically means and medians. For well-behaved distributions with finite moments up to 4, the mean will usually be the best estimator of location and it is the most natural for symmetric unimodal distributions like the normal distribution or the t-distribution with 3 or more degrees of freedom. However, for distributions such as the Cauchy, it is not defined, and in general, for highly skewed or heavy-tailed distributions, it is not the best measure of the "center of the distribution." For the Cauchy, the median and mode are equal and the sample median would be a good estimate of the center.

2.2.1   Estimating a Mean

For populations with a finite first moment, the mean is defined and can be estimated. When fourth moments exist, it is a stable estimate, and for symmetric unimodal distributions like the normal or the t with at least 3 degrees of freedom, it is a good (perhaps the best) measure of central tendency. How does the bootstrap enter into the picture of determining a point estimate of a population mean?

In the case of the Gaussian or even the highly skewed negative exponential distribution, the sample mean is the maximum likelihood estimate of the population mean; it is consistent and the minimum variance unbiased estimate of the population mean.

So how can the bootstrap top that? In fact, it cannot. We can determine the bootstrap estimate in this case without the need for a Monte Carlo approximation, and we would find that since the sample mean is the average over all bootstrap sample mean estimates, the bootstrap estimate is just equal to the sample mean. So it would be unnecessary to do the bootstrap, and especially to use a Monte Carlo approximation. Yet to establish theoretical properties for the bootstrap, the first proofs of consistency and asymptotic normality came for the bootstrap mean in the papers (see Singh, 1981 and Bickel and Freedman, 1981).

2.2.2   Estimating a Median

In situations where the population mean is not defined or the distribution is highly skewed, the median or mode (in the case of a unimodal distribution) would better represent the center of the distribution. Even for distributions like the Cauchy, the population median is well defined and the sample median is a consistent estimate of the population median. In such a case, the bootstrap estimate of the median is also consistent but again provides no advantage of the sample median. However, when we wish to estimate standard deviations for the sample estimates (the estimated standard error) of means or medians, the bootstrap will be much more useful.

2.3   ESTIMATING DISPERSION

When second moments exist, a standard deviation can be estimated. For unbiased estimates, this standard deviation is called the standard error of the estimate. Often, the standard deviation is considered the best measure of dispersion. This is because for normally distributed data and similar bell-shaped distributions, the empirical rule is based on the number of standard deviations an observation is away from the mean.

Also, for the class of distributions with finite second moments, Chebyshev's inequality puts a bound on the probability that an observation is k standard deviations away from the mean for k greater than 1. This helps us understand how the data are spread out.

However, for the Cauchy distribution, the tails are so heavy that the population mean does not exist and consequently neither does the variance. So the empirical rule and Chebyshev's inequality cannot be applied. In such instances, the interquartile range can be used as a measure of spread (variability).

2.3.1 Estimating an Estimate's Standard Error

In the text by Staudte and Sheather (1990, pp. 83–85), they provided an exact calculation for the bootstrap estimate of standard error for the median. This was originally derived by Maritz and Jarrett (1978). Staudte and Sheather then compared it with the Monte Carlo approximation to the bootstrap for a real sample of cell life data (obtained as the absolute value of the difference of seven pairs of independent identically distributed exponential random variables).

Following the approach of Staudte and Sheather, we shall derive their result for the median. For convenience, they assume that the sample size n is odd (so $n = 2m + 1$ for some integer m). This is done just to simplify the exposition and is not a critical assumption. Maritz and Jarrett (1978) actually provided explicit results for arbitrary n. We chose n odd so that the median is the unique "middle value" when n is odd but otherwise is the average of the two most central values when n is even. Let $X_{(i)}$ be the ith ordered observation. Then for odd n, the median is $X_{(m+1)}$.

They derive the exact variance of the bootstrap distribution of the median using well-known properties of order statistics. Let $X^*_{(1)}, X^*_{(2)}, \ldots, X^*_{(n)}$ be the order statistics of a bootstrap sample taken from $X_{(1)}, X_{(2)}, \ldots, X_{(n)}$. Let $X_{(i)}$ be the ith smallest observation in the original sample and $N^*_i = \#\{j : X^*_j = x_{(i)}, i = 1, 2, \ldots, n\}$. It can be shown that $\sum_{i=1}^{N} N^*_i$ has a binomial distribution with parameters n and p with $p = k/n$. Now, let P^* be the probability under bootstrap sampling. It follows that

$$P^*\left[X^*_{(m+1)} > x_k\right] = P^*\left\{\sum_{i=1}^{k} N^*_i \leq n\right\} = \sum_{j=0}^{n}\binom{n}{j}(k/n)^j\left[(n-k)/n\right]^{n-j}.$$

Using well-known relationships between binomial sums and incomplete beta functions, Staudte and Sheather (1990) found that

$$w_k = \left\{n!/(m!)^2\right\}\int_{(k-1)/n}^{k/n}(1-z)^m z^m dz,$$

and by simple probability calculations, the bootstrap variance of $X^*_{(m+1)}$ is

$$\sum_{i=1}^{n} w_k x_{(k)} - \left(\sum_{i=1}^{n} w_k x_{(k)}\right)^2.$$

The square root of the above expression is the bootstrap estimate of the standard deviation for $X^*_{(m+1)}$. Note that this is an explicit formula that does not require any Monte Carlo. For the application to the sister cell data, see Staudte and Sheather (1990, p. 85) or Chernick (2007, p. 50).

For other parameter estimation problems, the Monte Carlo approximation may be required. This is fairly straightforward. Let $\hat{\theta}$ be the sample estimate of θ and let $\hat{\theta}^*_i$ be the bootstrap estimate of θ based on the ith bootstrap sample and let θ^* be the average of the $\hat{\theta}_i^* s$. Then, as described in Efron (1982), the bootstrap estimate for the standard deviation for the estimator $\hat{\theta}$ is given by

$$SD_b = \left\{ [1/(k-1)] \sum_{i=1}^{n} \left(\hat{\theta}_i^* - \theta^* \right)^2 \right\}^{1/2}.$$

2.3.2 Estimating Interquartile Range

A natural way to estimate the interquartile range is to use the 25th percentile of the bootstrap distribution and the 75th percentile of the bootstrap distribution to obtain the estimate. If these cannot be done by exact computation, simply use the histogram of bootstrap sample estimates obtained by generating say 1000 bootstrap estimates ordering them from lowest to highest and use the 250th ordered bootstrap estimate and the 750th.

2.4 LINEAR REGRESSION

2.4.1 Overview

For the linear regression model, if the least squares estimation procedure is used to estimate the regression parameters and the model is reasonable and the noise term can be considered to be independent and identically distributed random variables with mean 0, and finite variance σ^2, bootstrapping will not add anything. That is because of the Gauss–Markov theorem that asserts that the least squares estimates of the regression parameters are unbiased and have minimum variance among all unbiased estimators. Under these assumptions, the covariance matrix for the least squares estimate $\hat{\beta}$ of the parameter vector β is given by $\Sigma = \sigma^2 (X^T X)^{-1}$, where X is called the design matrix and the inverse matrix $(X^T X)^{-1}$ exists as long as X has full rank. If $\hat{\sigma}^2$ is the least squares estimate of the residual variance, $\hat{\Sigma} = \hat{\sigma}^2 \left(X^T X \right)^{-1}$ is the commonly used estimate of the parameter covariance matrix.

Moreover, if the residuals can be assumed to have a Gaussian distribution, the least squares estimates have the nice additional property of being maximum likelihood estimates and are therefore the most efficient (accurate) estimates. If the reader is interested in a thorough treatment of linear regression, Draper and Smith (1981, 1998) are very good sources.

In the face of all these wonderful properties about the least squares estimates, you might ask why not just always use least squares? The answer is that you should not always use least squares and the reason is that the estimates are not robust. They can fail miserably when the modeling assumptions are violated. If the error distribution is heavy tailed or there are a few outliers in the data, the least squares estimates will give too much weight to the outliers, trying to fit the outliers well at the expense of the rest of the data.

Outliers have high influence on the regression parameters in that removing them would radically change the estimates. When the error distribution has heavy tails, robust procedures such as least absolute deviations, M-estimation, and repeated medians may be better approaches even though they are analytically more complex.

Regardless of the procedure used to estimate the regression parameters if we are interested in confidence regions for the parameters or want prediction intervals for future cases, we need to know more about the error distribution. When the Gaussian distribution is assumed, these confidence regions and prediction regions are straight-forward to compute. See Draper and Smith (1981, 1998) for details. However, when the error distribution is unknown and non-Gaussian, the bootstrap provides a way to get such estimates regardless of the method for estimating the parameters.

Other complications in regression modeling can also be handled by bootstrapping. These include heteroscedasticity of variances, nonlinearity in the model parameters, and bias due to transformation. Bootstrap approaches to the problem of heteroscedastic-ity can be found in Carroll and Ruppert (1988), and adjustment for retransformation bias can be found in Duan (1983). The works of Miller (1986, 1997) concentrated on linear models and are excellent sources for understanding the importance of the model-ing assumptions.

2.4.2 Bootstrapping Residuals

There are two basic approaches to bootstrapping in regression and both can be applied to nonlinear as well as linear regression problems. The first method is bootstrap-ping residuals. You apply a method such as least squares or least sum of absolute errors. The residual is then defined to be the observed value minus the regression esti-mate of it.

In theory, if the model is nearly correct, the model residuals would be independent and identically distributed. For each vector of prediction parameters and response, a residual can be computed. So then it is possible to sample the residuals with replace-ment, add the bootstrap residuals to the estimates to get a pseudosample, and then fit the model to the pseudosample and repeat. This method is easy to do and is often useful.

The following expression is a general form for a regression function and can apply to both linear and nonlinear regression:

$$Y_i = g_i(\beta) + \varepsilon_i \text{ for } i = 1, 2, \dots, n.$$

The function g_i is of a known form and is a function of a fixed set of covariates $\mathbf{C} = (C_1, C_2, \dots, C_p)$, β is the p-dimensional vector of coefficients associated with the covari-ates, and ε_i is independent and identically distributed from an unknown distribution F. We assume that F is "centered" at zero. By centered we usually mean that the expected value of ε_i is zero, but in general, we only require the median to be zero so as to allow for the case where the mean does not exist.

To estimate β based on observed data, we find the values that minimize a distance from $\lambda(\beta) = (g_1(\beta), g_2(\beta), \dots, g_n(\beta))^T$ to the observed $(y_1, y_2, \dots, y_n)$. In general, we denote the distance measure as $D(y, \lambda(\beta))$. As examples, the least squares criterion would be $D(y, \lambda(\beta)) = \sum_{i=1}^{n}(y_i - g_i(\beta))^2$, and for minimum absolute deviation $D(y, \lambda(\beta)) = \sum_{i=1}^{n}|y_i - g_i(\beta)|$.

Then, the estimate $\hat{\beta}$ is the value of β such that $D(y, \lambda(\hat{\beta})) = \min_{\beta} D(y, \lambda(\beta))$. The residuals are defined as $\hat{\varepsilon}_i = y_i - g_i(\hat{\beta})$ for $i = 1, 2, \dots, n$.

Now, to bootstrap the residuals, we take the empirical distribution for the residuals, which put probability $1/n$ for each $\hat{\varepsilon}_i$ for $i = 1, 2, \ldots, n$ and sample n times from this empirical distribution to get the bootstrap sample for the residuals, which is denoted as $\left(\varepsilon_1^*, \varepsilon_2^*, \ldots, \varepsilon_n^*\right)$. This then generates a bootstrap sample of observations $y_i^* = g_i\left(\hat{\beta}\right) + \varepsilon_i^*$ for $i = 1, 2, \ldots, n$.

Then for each bootstrap data set, we take the estimate β^* is the value of β such that

$$D(y, \lambda(\beta^*)) = \min_{\beta} D(y, \lambda(\beta)).$$

Using the usual Monte Carlo approximation, the process is repeated B times and the bootstrap estimate for the covariance matrix of $\hat{\beta}$ is the estimate:

$$\Sigma^* = (B-1)^{-1} \sum_{j=1}^{B} \left(\beta_j^* - \beta^*\right)\left(\beta_j^* - \beta^*\right)^T,$$

where β_j^* is the jth bootstrap estimate for $j = 1, 2, \ldots, B$ and $\beta^* = B^{-1} \sum_{j=1}^{B} \beta_j^*$. This is the covariance matrix estimate recommended by Efron (1982, p. 36).

For linear regression, $g_i(\beta)$ is the same function for each i and it is linear in the components of β. For nonlinear models, we show specific forms in Section 2.5.

2.4.3   Bootstrapping Pairs (Response and Predictor Vector)

In the linear model, when the error terms are normally distributed, bootstrapping residuals are fine. However, the method of bootstrapping vectors in practice is more robust to departures from normality and/or misspecification of the terms in the model. Bootstrapping vectors simply treats the vector that includes the response variable and the regression covariates as though they were independent and identically distributed random vectors (possibly with a correlation structure). But under these assumptions, it is easy to sample the vectors with replacement as the procedure for generating the bootstrap samples. Then, the bootstrap samples can be used as the data to generate the bootstrap estimates.

Efron and Tibshirani (1986) claimed that the two approaches are asymptotically equivalent when the model is "correct" but they can perform quite differently in small samples. Also, the bootstrapping vectors approach is less sensitive to the model assumptions and can still work reasonably when those assumptions are violated. This can be seen from the fact that the vector method does not make explicit use of the model structure while the residual method does.

2.4.4   Heteroscedasticity of Variance: The Wild Bootstrap

Wu (1986) discussed a jackknife approach to regression estimation, which he saw as superior to these bootstrap approaches. His approach has been shown to work particularly well when the residual variances are heteroscedastic. Bootstrapping vectors as described in this section (also called bootstrapping pairs in the econometric literature) can do a better job than bootstrapping residuals. Another approach to heteroscedasticity that bootstraps modified residuals is called the wild bootstrap. It was introduced in Liu (1988) based on suggestions by Wu and Beran in the discussion of Wu (1986). The wild bootstrap has several variants and has been shown to be effective in the heteroscedastic case and better than bootstrapping pairs. The variations in the wild bootstrap are based on the choice of a distribution of the variable by which to multiply the ordinary least squares residuals or a transformed version of them. For example, Liu (1988) suggested random multiplication by 1 or -1 with each having probability $1/2$ (which we call $F2$), while Mammen (1993) chose $-\left(\sqrt{5}-1\right)\big/2$ with probability $p=\left(\sqrt{5}+1\right)\big/\left(2\sqrt{5}\right)$ and $\left(\sqrt{5}+1\right)\big/2$ with probability $1-p$ (which is commonly used and called $F1$). The variable used for the multiplication is called an auxiliary variable. We cover this in more detail next.

The linear heteroscedastic model has the form

$$y_i = X_i\beta + u_i,$$

with $E(u_i|X_i) = 0$ and $E\left(u_i^2\big|X_i\right) = \sigma_i^2$.

What distinguishes this from the standard linear model is the dependence of the variance of the error term u_i on X_i. This problem can be solved if the variance σ_i^2 has a specific form. If the form of σ_i^2 is completely unknown, then we have no way of mimicking it in the process used to draw the bootstrap samples. When we have this situation, the wild bootstrap comes to the rescue. Both Liu (1988) and Mammen (1993) showed, under a variety of regularity conditions, that the wild bootstrap is asymptotically consistent. For hypothesis testing, Davidson and Flachaire (2001, 2008) demonstrated that small sample size tests based on $F2$ do better than those based on the popular $F1$, which is commonly preferred because it corrects for skewness.

In Section 3.6, we will look at simulation results for constructing bootstrap confidence intervals for variances that the more complicated bootstraps with the higher-order asymptotic accuracy do not necessarily outperform or even perform as well as the simpler bootstrap confidence intervals when the sample size is small. It is the effect of estimating adjustment terms from the data that adds variability to the procedure that is significant when the sample size is small. The same thing might hold here if the adjustment was based on the sample as the sample estimate would add to the variability. But this is not the case. However, since the adjustment is right for a very specific skewness, it may be biased as the level of skewness could be more or less extreme depending on the actual sampling distribution.

In addition to the work of Davidson and Flachaire (2001), the simulation results of Godfrey and Orme (2001) indicate that $F2$ is the better procedure. The early work of Freedman (1981) demonstrated that bootstrapping pairs was better than bootstrapping residuals under almost any form of model misspecification. However, some simulation work by Horowitz (2000) indicates that bootstrapping pairs is not always accurate. But this deficiency was corrected by a new version of the pairs bootstrapping for hypothesis testing that was given by Mammen (1993) and subsequently modified by Flachaire (1999). See Flachaire (2002) for a more detailed account of the advantages of wild bootstrap over bootstrapping pairs.

Shao (2010) extended the concept of wild bootstrap to dependent data and called it the dependent wild bootstrap. As this is an alternative to block bootstrapping in time series data, we will discuss it in detail in Section 5.6.

2.4.5   A Special Class of Linear Regression Models: Multivariable Fractional Polynomials

The multivariable fractional polynomial (MFP) model was proposed by Sauerbrei and Royston (1999) as a method for combining variable selection with the identification of a functional form (linear vs. nonlinear) for the continuous predictor variables included in the model. The method works in a very systematic way. The model considers terms that are models that are called $FP1$ and $FP2$. $FP1$ functions have the form $\beta_1 Xp1$ where β_1 is the regression parameter and $p1$ is a power chosen from the set $S = (-2, -1, -0.5, 0, 0.5, 1, 2, 3)$. $p1 = 0$ simply denotes $\log X$ (see Royston and Altman, 1994). An $FP2$ function is simply a function of the form

$$\beta_1 X^{p1} + \beta_2 X^{p2},$$

where $p1 \neq p2$, and both are chosen from S. We also include repeated powers or $p1 = p2$, but then the form is

$$\beta_1 X^{p1} + \beta_2 X^{p1} \log X.$$

There are then 8 possible $FP1$ functions including the linear ($p1 = 1$) and 64 $FP2$ functions. A method for selecting the best fitting of these functions is described in Royston and Sauerbrei (2005).

These models have been found to be very useful in identifying good prognostic factors from clinical trial data (Sauerbrei and Royston, 1999). However, they acknowledge that the stability of the model must be checked. This is what they did in Royston and Sauerbrei (2003). They bootstrapped the entire model building procedure by taking 5000 bootstrap samples from the patient population.

We will see in Chapter 7 that this idea was first tried by Gail Gong in her PhD thesis. Also, the idea of finding prognostic factors that can show qualitative interactions with treatments is the same idea that Lacey Gunter used for her PhD thesis. This will also be covered in Chapter 7.

Sauerbrei and Royston (2007) pointed out that the bootstrap is underutilized in controlled clinical trials, and they argued by example how the bootstrap could play a bigger role in such trials. In that article, they also sketched the results from some of their earlier papers.

2.5   NONLINEAR REGRESSION

The theory of nonlinear regression advanced significantly in the decades of the 1970s and 1980s. Much of this development has been documented in textbooks devoted entirely to nonlinear regression models. Two prime examples are Bates and Watts (1988) and Gallant (1987).

The nonlinear models can be broken down into two categories. The first category of models is for the ones that allow for reasonable local linear approximations through the Taylor series expansions. When this can be done, approximate confidence and/or prediction intervals can be generated using asymptotic theory. Much of this theory has been covered by Gallant (1987).

In the aerospace industry, orbit determination is a nonlinear estimation problem, which is often successfully tackled by linear approximations (i.e., the Kalman–Bucy filter). Kalman filters are used to dynamically determine positions of objects orbiting earth such as communication (Milstar) and navigational (global positioning system [GPS]) satellites as well as tracking the trajectories of incoming ballistic missiles. The recursive formulae shows the position of the object at a later time given its position one time step earlier. The second category is the highly nonlinear class of models for which the linear approximation will not work. Bates and Watts (1988) provided methods for diagnosing the degree of nonlinearity, which allows users to determine which category their model falls into.

As Efron (1982) pointed out, the bootstrap can be applied to just about any nonlinear problem. To bootstrap, you do not need to have differentiable functional forms. The model can even be described by a complicated computer algorithm rather than a simple analytical expression. The residuals do not need to have a Gaussian distribution. The only restriction on the residuals is that the residuals at least be exchangeable (independent and identically distributed would be a special case of exchangeability) and have a finite variance. Recall that sometimes certain regularity conditions are also needed to guarantee consistency of the bootstrap estimates.

2.5.1 Examples of Nonlinear Models

Many regression problems can be solved using models that are linear in the parameters or can be made linear by transformation. In regression modeling, the difference between linear and nonlinear models is in the form of the model as linear or nonlinear functions of the parameters and not the form of the model as a function of the variables. Hence, the model

$$f(x, \theta) = \theta_1 + \theta_2 e^x + \varepsilon$$

is linear because it is a linear function of $\theta = (\theta_1, \theta_2)^T$, even though it is a nonlinear function of x. On the other hand, the function

$$g(x, \theta) = \theta_1 + \theta_2 e^{\theta_3 x} + \varepsilon$$

is nonlinear because the term $e^{\theta_3 x}$ is a nonlinear function of θ_3. The function g given above could arise as the solution to a simple differential equation where x is given and g is determined with measurement error.

A common problem in time series analysis is called the harmonic regression problem. We may know that the response function is periodic or the sum of a few periodic functions. However, we do not know the amplitudes or frequencies of these components. The fact that the frequencies are unknown parameters is what makes this a nonlinear problem. We now illustrate this with a simple function with just one periodic component. We let

$$f(t, \varphi) = \varphi_0 + \varphi_1 \sin(\varphi_2 t + \varphi_3) + \varepsilon,$$

where t is the time.

Because of the trigonometric identity $\sin(A + B) = \sin A \cos B + \cos A \sin B$, the term $\varphi_1 \sin(\varphi_2 t + \varphi_3)$ can be reexpressed as $\varphi_1 \cos \varphi_3 \sin(\varphi_2 t) + \varphi_1 \sin \varphi_3 \cos(\varphi_2 t)$.

The reparameterization could be expressed as

$$f(t, A) = A_0 + A_1 \sin A_2 t + A_3 \cos A_2 t,$$

where

$$A = (A_0, A_1, A_2, A_3)^T$$

and

$$A_0 = \varphi_0,$$
$$A_1 = \varphi_1 \cos \varphi_3,$$
$$A_2 = \varphi_2,$$

and

$$A_3 = \varphi_1 \sin \varphi_3.$$

This reparameterization was done to illustrate the model in the form given in Gallant (1987).

There are many other examples where nonlinear models are solutions of linear differential equations or systems of differential equations. The solutions involve exponential functions (both real and complex valued). Then, the resulting functions are real valued and exponential or periodic or a combination of both. As a simple example, consider the equation

$$dy(x)/dx = -\varphi_1 y(x)$$

subject to the initial condition $y(0) = 1$. The solution is then

$$y(x) = e^{-\varphi_1 x}.$$

Now, since the parameter of the solution is the unknown φ_1 and $y(x)$ is assumed to be observed with additive noise, the equation produces a nonlinear regression model:

$$y(x) = e(-\varphi_1 x) + \varepsilon$$

For a commonly used linear system of differential equations whose solution involves a nonlinear model, see Gallant (1987, pp. 5–8). This type of problem occurs frequently when compartmental models are used, which is particularly the case for chemical kinetics problems.

2.5.2 A Quasi-Optical Experiment

In this experiment, Shimabukuro et al. (1984) applied the bootstrap to a nonlinear model in order to determine the precision of estimates of quasi-optical characteristics of materials. These experimenters wanted good estimates of standard errors for the parameter estimates so that they could assess the accuracy of their new measurement technique to the ones that were previously used and have published estimates.

The parameters are called the loss tangent and the dielectric constant (or relative permittivity) and are related to the transmission of millimeter wavelength signals through a dielectric slab. The model involves the following equations but was simply given as a computer program to which the bootstrap was applied to generate bootstrap samples of the measurement data.

Measurements are taken to compute $|T|^2$, the squared norm of a complex number T known as the transmission coefficient. An expression for T is

$$T = \left(1 - r^2\right) e^{-(\beta_1 - \beta_0)di} \big/ \left(1 - r^2 e^{-2\beta_1 di}\right),$$

where

$$\beta_1 = 2\pi \left(\sqrt{\varepsilon_1/\varepsilon_0} - \sin^2 \varphi\right) \big/ \lambda_0,$$
$$\beta_0 = 2\pi \cos \varphi / \lambda_0,$$
$$\varepsilon_1 = \varepsilon_r \varepsilon_0 \left\{1 - i\sigma/(\omega \varepsilon_r \varepsilon_0)\right\},$$

and

ε_0 = permittivity of free space, ε_r = relative permittivity, σ = conductivity, λ_0 = free-space wavelength, $\tan \delta = \sigma/(\omega \varepsilon_r \varepsilon_0)$ = loss tangent, d = thickness of the slab, r = reflection coefficient of a plane wave incident to a dielectric boundary, ω = free-space frequency, and $i = \sqrt{-1}$.

For more details on the results and the experimental setup, refer to Chernick (2007), and for a more complete account of the experiment, refer to Shimabukuro et al. (1984).

2.6 NONPARAMETRIC REGRESSION

Given a vector X, the regression function $E(y|X)$ is often a smooth function of X. In Sections 2.4 and 2.5, we considered linear and nonlinear parametric models for the regression function. In nonparametric regression, we allow the regression function to come from a general class of smooth functions.

The nonparametric regression function is given by the following equation when X is one-dimensional, and we observe (y_i, x_i) for $i = 1, 2, \ldots, n$:

$$y_i = g(x_i) + \varepsilon_i,$$

where $g(x) = E(y|x)$ is the function we wish to estimate. We assume that the ε_i is independent and identically distributed (i.i.d.) with mean zero and finite constant variance σ^2. One approach to the estimation of g is kernel smoothing (see Hardle, 1990a, b or Hall, 1992, pp. 257–269]). One application of the bootstrap here is to determine the degree of smoothing to use (i.e., determine the trade-off between bias and variance as is also done in nonparametric kernel density estimation).

In a very recent paper, Loh and Jang (2010) took a novel bootstrap approach to bandwidth selection for a two-point correlation function of a spatial point process. Instead of using the bootstrap to estimate the variance and the bias, the authors chose a nonparametric spatial bootstrap procedure to estimate the variance's component and a different method to estimate the bias component (a plug-in estimator based on a parametric model). See the paper for the details and the reason for choosing this method over simply bootstrapping to estimate the entire integrated squared error.

2.6.1   Examples of Nonparametric Regression Models

2.6.1.1   Cox Proportional Hazard Model and Kaplan–Meier Method (a Semiparametric and a Nonparametric Survival Method).   An approach that we include here is the Cox proportional hazard model (Cox, 1972). It is actually considered to be a semiparametric method, and it looks like a form of nonlinear regression. This type of model is very common in the analysis of clinical trials where survival or some other time to event point is considered, and right censoring occurs because many of the subjects will not have an event during the course of the trial, which could involve 1–2 years of follow-up. The hazard rate function, denoted as $h(t|x)$, is the first derivative of the survival function $S(t|x)$ = the probability of surviving t or more units of time given the predictor vector or covariate vector x.

The Cox model is then $h(t|x) = h_0(t)e^{\beta x}$, where $h_0(t)$ is an arbitrary unspecified baseline hazard rate function assumed to depend solely on t. This equation shows the proportional hazard rate assumption. The hazard rate $h(t|x)$ is proportional to the baseline hazard rate at time t with the proportionality given by $e^{\beta x}$. Through the use of the "partial likelihood" function that Cox introduced, the parameter vector β can be estimated by maximizing this partial likelihood function. Because of the existence of the underlying hazard rate function $h_0(t)$, the method is not fully parametric.

Efron and Tibshirani (1986) applied the bootstrap for leukemia data in mice to access the efficacy of a treatment on their survival. See the article for the details. In addition to the standard Kaplan–Meier curve and the standard error approximations by the Peto method or the Greenwood method (see Chernick and Friis, 2003, chapter 15), the bootstrap can be used to obtain standard error estimates.

These methods are the most commonly used semiparametric and nonparametric approaches to survival analysis with right-censored data. Efron (1981) compared Greenwood's formula with bootstrap estimation for a particular data set called the Channing House Data. This example is discussed in more detail in Section 8.4.

Recently, Liu et al. (2010) produced improved estimates of the regression parameters in a Cox model applied to a nested case-control design with auxiliary covariates. They demonstrate the superiority of their estimate over the Thomas estimator (Thomas 1977) through simulation and illustrate the technique on a real data example. Proposition 1 in the paper asserts the asymptotic normality of their estimator under certain conditions including some regularity conditions. The asymptotic covariance matrix for the estimator has a component they denote as Ω. Because the estimation of the matrix Ω is complicated, the authors applied the bootstrap to obtain a consistent estimate of Ω. See Liu et al. (2010) for details about the estimator and the structure of Ω.

2.6.1.2 Alternating Conditional Expectation (ACE) and Projection Pursuit.
Two other nonparametric regression techniques are ACE and projection pursuit. We shall not go into a description of these techniques but mention them because as with many complex models where it is difficult to estimate standard errors or compute confidence intervals, the bootstrap can play a vital role. See Efron and Tibshirani (1986) for applications of the bootstrap when these two modeling procedures are used.

2.6.1.3 Classification and Regression Trees.
Regression trees date back to the work at the University of Michigan's Institute for Social Research in the 1960s. Their algorithm was called AID for Automatic Interaction Detection. Later in the 1970s, the same institute developed a tree classifier called Tree Hierarchical Automatic Interaction Detection (THAID). However, in the mid-1970s, Breiman and Friedman started working on this idea independently and without knowledge of the previous work done at Michigan.

The main reason that tree classification methods had not developed during the 1960s was that the early methods that fit the data well were often poor predictor algorithms. Growing trees was not a difficult problem, but constructing them in a way that avoided overfitting was a major stumbling block.

But by the early 1980s, Breiman et al. solved the problem by growing an overly large tree and then figuring the right way to prune back some of the branches. This was a breakthrough that made classification trees competitive with other standard classification rules. They used their methods in their consulting work and found it particularly useful in the medical field.

The text "Classification and Regression Trees" appeared in 1984 as well as the commercial software CART. As these professors were not adept at producing good front end software, they eventually sold the rights to Salford Systems. In their text, Breiman et al. (1984) provided very simple data examples to show how easily interpretable these classification trees could be.

One example was a method to do automatic pattern recognition for the digits 0, 1, 2, 3, 4, 5, 6, 7, 8, and 9. The initial and most striking example for its simplicity is the rule to predict risk of death within 30 days for patients admitted to the University of

California at San Diego Medical Center after experiencing a heart attack. This was to be a two-class problem where the patient is classified as high risk or not high risk.

G was used to denote high risk and F to indicate not at high risk. The patients have 19 vital sign measurements taken in the first 24 h. These were blood pressure, age, and 17 other ordered categorical and binary variables that the physicians believed character-ized the patient's condition. Olshen constructed, as an example in the book, a very effective rule using the CART algorithm based solely on three variables systolic blood pressure, age, and the presence of sinus tachycardia. The rule goes as follows:

Step 1. Determine if the minimum systolic blood pressure over the first 24 h is above 91. If not, then the patient is at high risk.

Step 2. If at step 1 the answer is yes, then check the age of the patient. If the patient is under 62.5 years of age, he or she is low risk.

Step 3. If the patient gets to step 2 and is over 62.5 years, determine whether or not he or she has sinus tachycardia. At this point, we declare the patients with sinus tachycardia as high risk, and the remaining others are low risk. This simple rule turned out to be very accurate and made sense medically.

Cross-validation methods (more general than leave-one-out) were used to get "honest" estimates of the predictive ability of these tree classifiers. The bootstrap was an alterna-tive but was not preferred by the developers of CART. Now, a later breakthrough that created a large improvement over CART was the construction of several trees that were used to produce a combination rule. The procedures of bagging and boosting were methods that combined the information from several trees to produce a better algorithm (e.g., use the majority rule for each vector of observed characteristics). Boosting and bagging fall under a class of methods that are now generally referred to as ensemble methods.

Recently, Chipman et al. (2010) used a Bayesian approach and applied it to regres-sion trees by summing the values of the approximating function to $E(Y|x)$ with $h(x) = \sum g_i(x)$ with each $g_i(x)$ being a regression tree. The choice of the m regression trees to select for the sum is based on a Bayesian learning method and the methodology is called BART (because it is a Bayesian additive regression tree procedure and the name sounds similar to CART).

Breiman (1996) developed a bootstrap approach called bagging, and he called the method applied to classification and regression trees random forests. We will look at bagging in detail in the next section.

2.6.2 Bootstrap Bagging

Advances in data mining, statistical inference, prediction, and classification are taking place in several disciplines. Computer science and engineering have been the disci-plines leading the way, but the statistical profession is rapidly finding its appropriate place. The text from Hastie et al. (2009) is a comprehensive source for these develop-ments from the statistical point of view. Although in life we find that committee decisions are often better than individual decisions, it is only very recently that this has been applied to machine learning. This began with the AdaBoost algorithm of Freund and Schapire (1997) and related techniques now referred to as boosting. This idea started in the supervised learning (classification) arena and was shockingly successful there. Other methods similar in spirit are bagging (Breiman, 1996) and random forests (Breiman, 2001).

In regression or classification, the prediction rules can be improved by reducing variability of the estimate through a technique called bootstrap bagging (Breiman, 1996). First, let us consider a simple regression problem. We wish to fit a model to our data:

$$X = \{(u_1, y_1), (u_2, y_2), \ldots, (u_n, y_n)\}.$$

Our regression function f is then estimated from X and evaluated for prediction of y at any u in the domain of f. Bagging, also referred to as bootstrap aggregation, generates copies of f by resampling the n pairs (u_i, y_i), $i = 1, 2, \ldots, n$ at random and with replacement n times. Using the Monte Carlo approximation to the bootstrap, we generate B bootstrap samples of size n, and for each one, we compute f by the regression method we used on the original sample. The resulting functions we denote as f^{*b}. The bagged estimate of y at any u is given by

$$f_{\text{bag}}(u) = \left[\sum_{b=1}^{B} f^{*b}(u) \right] \Big/ B.$$

This is simply the average of the bootstrap estimate over all the bootstrap samples. This estimate will have the same expected value as the f we got from the original sample, but because of the averaging, it will have a smaller variance. Hence, it provides a better prediction of y given the value u. This idea can be applied to any regression method including linear, nonlinear, and nonparametric regression methods covered earlier in this chapter.

However, it is particularly useful for the regression tree. It works best when the prediction rule is highly variable, which can be the case with regression or classification trees. The same idea can be applied to classification where averaging is replaced by majority rule. Hastie et al. (2009, pp. 282–288) provided a more detailed treatment, some interesting examples, and nice color graphics.

2.7 HISTORICAL NOTES

For the error rate estimation problem, there is currently a large body of literature. Developments up to 1974 are well covered in Kanal (1974), and Toussaint (1974) is an extensive bibliography on the topic. For multivariate Gaussian feature vectors, McLachlan (1976) showed that the asymptotic bias of resubstitution is greater than zero, indicating that resubstitution is not always a consistent estimator of the true error rate. The bias of plug-in estimates of error rates is discussed in Hills (1966).

Choi (1986) is a collection of articles on classification and includes some of the work on bootstrap. Most of the simulation studies consider only linear discriminant functions, although Jain et al. (1987) looked at quadratic discriminant rules and the performance of the various resampling estimators including the 632 estimator. They also considered only the two-class problem with two-dimensional features.

But Efron (1982, 1983) and Chernick et al. (1985, 1986, 1988a) also included five-dimensional feature vectors in some of their simulations. Chernick, Murthy, and Nealy also considered three-class problems in some cases. Chernick et al. (1988a, b) were the first to look at results for non-Gaussian distributions such as multivariate Cauchy and Pearson VII elliptically contoured distributions.

Hirst (1996) developed a smoothed estimator, extending the work of Snapinn and Knoke (1984, 1985a) and Glick (1978). Snapinn and Knoke (1984, 1985a) compared their method to cross-validation but unfortunately no comparison with bootstrap methods. Hirst (1996) was the first to seriously compare his smoothed estimates to bootstrap 632.

Confidence intervals as well as standard errors were studied in the simulations of Chatterjee and Chatterjee (1983). Jain et al. (1987) were the first ones to look at non-Gaussian populations and the characteristics of quadratic and nearest neighbor classi-fiers. Chernick et al. (1988a, b), Hirst (1996), and Snapinn and Knoke (1985b) considered various types of non-Gaussian distributions.

Efron and Tibshirani (1997) generalized 632 to form an even better estimate that they call 632+. Bagging classification trees was introduced by Breiman (1996) and random forests by Breiman (2001). The breakthrough with boosting came in the Freund and Schapire (1997) article. Fu et al. (2005) showed that bootstrap cross-valdation has advantages over 632 and 632+ in several examples.

Another good survey article is McLachlan (1986). McLachlan (1992) provided extensive coverage on the classification problem and error rates. The Pearson VII family of distributions is elliptically contoured like the Gaussian as specified in the parameter space but can have heavier or lighter tails than the Gaussian. In the heavy-tailed cases, the superiority of the 632 estimator disappears. This family of distributions has its tails governed by the parameter M (see Johnson, 1987).

The bootstrap distribution for the median is also discussed in Efron (1982, pp. 77–78) and Mooney and Duval (1993). Consistency of the bootstrap estimate in a bioequivalence example is given by Shao et al. (2000).

Early work on bootstrapping in regression was Freedman (1981) and was followed by Bickel and Freedman (1983), Weber (1984), Wu (1986), and Shao (1988a, b). There are many fine books on regression analysis including Draper and Smith (1981) for linear regression and Gallant (1987) and Bates and Watts (1988) for nonlinear regression. But none of these discuss the bootstrap approach. Sen and Srivastava (1990) is a more recent text that does cover bootstrapping somewhat. There is a general trend to include some discussion of the bootstrap in most advanced and even some elementary statistics texts these days. Another example where the bootstrap is at least introduced is the more recent edition (Draper and Smith, 1998). Efron (1982) discussed early on the bootstrap-ping pairs and bootstrapping residuals approaches in nonlinear regression and the heteroscedastic linear regression model.

Efron and Tibshirani (1986) provided a number of interesting applications for the bootstrap in regression problems including nonparametric regression approaches such as projection pursuit regression and methods for deciding on transformations of the response variable such as the ACE method due to Breiman and Friedman (1985). The work of Shimabukuro et al. (1984) was an early practical application of bootstrapping in nonlinear regression. Daggett and Freedman (1985) applied the bootstrap to an econometric model.

There have also been applications by Breiman (1992) for model selection related to x-fixed prediction, Brownstone (1992) regarding admissibility of linear model selection techniques, Bollen and Stine (1993) regarding structural equation modeling, and Cao-Abad (1991) regarding rates of convergence for the wild bootstrap. DeAngelis et al. (1993) applied the bootstrap to L1 regression. Lahiri (1994) applied the bootstrap using M-estimation. Dikta (1990) applied the bootstrap to nearest-neighbor regression and Green et al. (1987) to the estimation of price elasticities. Shi (1991) introduced the local bootstrap that can handle heteroscedasticity for both linear and nonparametric regression.

2.8   EXERCISES

1. Airline accidents: According to the U.S. National Transportation Safety Board, the number of airline accidents by year from 1983 to 2006 were 23, 16, 21, 24, 34, 30, 28, 24, 26, 18, 23, 23, 36, 37, 49, 50, 51, 56, 46, 41, 54, 30, 40, and 31.

   (a) For the sample data, compute the mean and its standard error (from the standard deviation), and the median.

   (b) Using R, compute bootstrap estimates of the mean and median with estimates of their standard errors, using $B = 1000$ resamples. Also, compute the median of the median estimates. (Use set.seed(1) to allow your results to be reproduced.)

   (c) Compare Exercise 2.8.1a,b. How do the estimates compare?

   Hint: Use the following R script:

```
set.seed(1)
x<- c(23, 16, 21, 24, 34, 30, 28, 24, 26, 18, 23, 23, 36,
37, 49, 50, 51, 56, 46, 41, 54, 30, 40, 31)
n<- length(x)
B<- 1000
xMean<- NULL
xMed<- NULL
xSEmean<- NULL
xSEmed<- NULL
for (i in 1:B) { #for each bootstrap resample
xx<- sample(x, n, replace=TRUE) #resample
```

```
xMean[i]<- mean(xx) #keep track of mean estimates
xMed[i]<- median(xx) #keep track of median estimates
}
c(mean(x), sd(x)/sqrt(n)) #show mean of data and calculated
standard error
c(mean(xMean), sd(xMean)) #show mean estimate and standard
error
median(x) #show median of data
c(mean(xMed), sd(xMed)) #show median estimate and standard
error
median(xMed) #show median of medians
```

2. Using the aflatoxin data of Exercise 1.6.5, and the R script of Exercise 2.8.1 above (but change the data used for "x"):

   (a) For the sample data, compute the mean and its standard error (from the standard deviation), and the median.

   (b) Using R, compute bootstrap estimates of the mean and median with estimates of their standard errors, using $B = 1000$ resamples. Also, compute the median of the median estimates. (Use set.seed(1) to allow your results to be reproduced.)

   (c) Compare Exercise 2.8.2a,b. How do the estimates compare?

3. In Section 2.4.2, the method of "bootstrapping residuals" was described. In Section 2.4.3, the method of "bootstrapping pairs" or "bootstrapping vectors" was described. The latter might also be described as "case-wise" or "list-wise" or "row-wise" bootstrapping.

   (a) "Bootstrapping residuals" are conditional on the model used and the method of fitting. (These are held fixed under bootstrapping and must be known in advance.) Residuals in model fits are typically not independent and are correlated. (Errors are typically assumed independent, but their estimates as residuals are not.)

      (1) How would this affect the accuracy of, say, estimated standard errors of parameters fitted?

      (2) In binary logistic regression, the responses correspond to either "0" or "1" values. Would there be a problem bootstrapping by residuals in this situation?

   (b) "Bootstrapping vectors" are unconditional (does not require foreknowledge of modeling) and can be applied to any set of data in the form of replicate rows for columns corresponding to covariates. This includes multivariate data where there is no identified dependent response variable.

      (1) How does bootstrapping vectors relate to bootstrap bagging of Section 2.6.2?

      (2) Is there any difficulty in using bootstrapping vectors in the binary logistic regression problem of Exercise 2.8.3a, number 2?

REFERENCES

Ambroise, C., and McLachlan, G. J. (2002). Selection bias in gene extraction on the basis of microarray gene-expression data. Proc. Natl. Acad. Sci. USA 99, 6562–6566.

Bates, D. M., and Watts, D. G. (1988). Nonlinear Regression Analysis and Its Applications. Wiley, New York.

Bickel, P. J., and Freedman, D. A. (1981). Some asymptotic theory for the bootstrap. Ann. Stat. 9, 1196–1217.

Bickel, P. J., and Freedman, D. A. (1983). Bootstrapping regression models with many parameters. In Festschrift for Erich Lehmann (P. J. Bickel, K. Doksum, and J. L. Hodges, eds.), pp. 28–48. Wadsworth, Belmont, CA.

Bollen, K. A., and Stine, R. A. (1993). Bootstrapping goodness-of-fit measures in structural equation models. In Testing Structural Equation Models (K. A. Bollen, and J. S. Long, eds.), pp. 111–115. Sage Publications, Beverly Hills, CA.

Braga-Neto, U. M., and Dougherty, E. R. (2004). Is cross-validation valid for small-sample microarray classification? Bioinformatics 20, 374–380.

Breiman, L. (1992). The little bootstrap and other methods of dimensionality selection in regression, x-fixed prediction error. J. Am. Stat. Assoc. 87, 738–754.

Breiman, L. (1996). Bagging predictors. Mach. Learn. 26, 123–140.

Breiman, L. (2001). Random forests. Mach. Learn. 45, 5–32.

Breiman, L., and Friedman, J. H. (1985). Estimating optimal transformations for multiple regression and correlation. J. Am. Stat. Assoc. 80, 580–619.

Breiman, L., Friedman, J., Olshen, R., and Stone, C. (1984). Classification and Regression Trees. Wadsworth, New York.

Brownstone, D. (1992). Bootstrapping admissible linear model selection procedures. In Exploring the Limits of Bootstrap (R. LePage, and L. Billard, eds.), pp. 327–344. Wiley, New York.

Cao-Abad, R. (1991). Rate of convergence for the wild bootstrap in nonparametric regression. Ann. Stat. 19, 2226–2231.

Carroll, R. J., and Ruppert, D. (1988). Transformations and Weighting in Regression. Chapman & Hall, New York.

Chatterjee, S., and Chatterjee, S. (1983). Estimation of misclassification probabilities by bootstrap methods. Commun. Stat. Simul. Comput. 12, 645–656.

Chernick, M. R. (2007). Bootstrap Methods: A Guide for Practitioners and Researchers, 2nd Ed. Wiley, Hoboken, NJ.

Chernick, M. R., and Friis, R. H. (2003). Introductory Biostatistics for the Health Sciences: Modern Applications Including Bootstrap. Wiley, Hoboken, NJ.

Chernick, M. R., Murthy, V. K., and Nealy, C. D. (1985). Applications of bootstrap and other resampling techniques: Evaluation of classifier performance. Pattern Recognit. Lett. 3, 167–178.

Chernick, M. R., Murthy, V. K., and Nealy, C. D. (1986). Correction note to applications of bootstrap and other resampling techniques: Evaluation of classifier performance. Pattern Recognit. Lett. 4, 133–142.

Chernick, M. R., Murthy, V. K., and Nealy, C. D. (1988a). Estimation of error rate for linear discriminant functions by resampling: Non-Gaussian populations. Comput. Math. Appl. 15, 29–37.

Chernick, M. R., Murthy, V. K., and Nealy, C. D. (1988b). Resampling-type error rate estimation for linear discriminant functions: Pearson VII distributions. Comput. Math. Appl. 15, 897–902.

Chipman, H. A., George, E. I., and McCulloch, R. E. (2010). BART: Bayesian additive regression trees. Ann. Appl. Stat. 4, 266–298.

Choi, S. C. (1986). Discrimination and classification: Overview. Comput. Math. Appl. 12A, 173–177.

Cox, D. R. (1972). Regression models and life tables. J. R. Stat. Soc. B 34, 187–202.

Daggett, R. S., and Freedman, D. A. (1985). Econometrics and the law: A case study in the proof of antitrust damages. Proc. Berkeley Symp. VI, 123–172.

Davidson, R., and Flachaire, E. (2001). The wild bootstrap, tamed at last. Working paper IERR 1000, Queen's University, Ontario.

Davidson, R., and Flachaire, E. (2008). The wild bootstrap, tamed at last. J. Econom. 146, 162–169.

DeAngelis, D., Hall, P., and Young, G. A. (1993). Analytical and bootstrap approximations to estimator distributions in L1 regression. J. Am. Stat. Assoc. 88, 1310–1316.

Dikta, G. (1990). Bootstrap approximation of nearest neighbor regression function estimates. J. Multivar. Anal. 32, 213–229.

Draper, N. R., and Smith, H. (1981). Applied Regression Analysis, 2nd Ed. Wiley, New York.

Draper, N. R., and Smith, H. (1998). Applied Regression Analysis, 3rd Ed. Wiley, New York.

Duan, N. (1983). Smearing estimate a nonparametric retransformation method. J. Am. Stat. Assoc. 78, 605–610.

Duda, R. O., and Hart, P. E. (1973). Pattern Recognition and Scene Analysis. Wiley, New York.

Dudoit, S., Fridlyand, J., and Speed, T. P. (2002). Comparison of discrimination methods for the classification of tumors using gene-expression data. J. Am. Stat. Assoc. 97, 77–87.

Efron, B. (1981). Censored data and the bootstrap. J. Am. Stat. Assoc. 76, 312–319.

Efron, B. (1982). The Jackknife, the Bootstrap and Other Resampling Plans. SIAM, Philadelphia.

Efron, B. (1983). Estimating the error rate of a prediction rule: Improvements on cross-validation. J. Am. Stat. Assoc. 78, 316–331.

Efron, B., and Tibshirani, R. (1986). Bootstrap methods for standard errors, confidence intervals and other measures of statistical accuracy. Stat. Sci. 1, 54–77.

Efron, B., and Tibshirani, R. (1994). An Introduction to the Bootstrap. Chapman & Hall, New York.

Efron, B., and Tibshirani, R. (1997). Improvements on cross-validation: The 632+ rule. J. Am. Stat. Assoc. 92, 548–560.

Ferguson, T. S. (1967). Mathematical Statistics: A Decision Theoretic Approach. Academic Press, New York.

Flachaire, E. (1999). A better way to bootstrap pairs. Econ. Lett. 64, 257–262.

Flachaire, E. (2002). Bootstrapping heteroskedasticity consistent covariance matrix estimator. Comput. Statist. 17, 501–506.

Freedman, D. A. (1981). Bootstrapping regression models. Ann. Stat. 9, 1218–1228.

Freund, Y., and Schapire, R. (1997). A decision-theoretic generalization of online learning and an application to boosting. J. Comput. Sys. Sci. 55, 119–139.

Fu, W. J., Carroll, R. J., and Wang, S. (2005). Estimating misclassification error with small samples via bootstrap cross-validation. Bioinformatics 21, 1979–1986.

Fukunaga, K. (1990). Introduction to Statistical Pattern Recognition, 2nd Ed. Academic Press, San Diego, CA.

Gallant, A. R. (1987). Nonlinear Statistical Models. Wiley, New York.

Geisser, S. (1975). The predictive sample reuse method with applications. J. Am. Stat. Assoc. 70, 320–328.

Glick, N. (1978). Additive estimators for probabilities of correct selection. Pattern Recognit. 10, 211–222.

Godfrey, L. G., and Orme, C. D. (2001). Significance levels of heteroskedasticity-robust tests for specification and misspecification: Some results on the use of wild bootstraps. Paper presented at ESEM'2001, Lausanne.

Green, R., Hahn, W., and Rocke, D. (1987). Standard errors for elasticities: A comparison of bootstrap and asymptotic standard errors. J. Bus. Econ. Stat. 5, 145–149.

Hall, P. (1992). The Bootstrap and Edgeworth Expansion. Springer-Verlag, New York.

Hand, D. J. (1981). Discrimination and Classification. Wiley, Chichester.

Hand, D. J. (1982). Kernel Discriminant Analysis. Wiley, Chichester.

Hardle, W. (1990a). Applied Nonparametric Regression. Cambridge University Press, Cambridge.

Hardle, W. (1990b). Smoothing Techniques with Implementation in S. Springer-Verlag, New York.

Hastie, T., Tibshirani, R., and Friedman, J. (2009). The Elements of Statistical Learning: Data Mining, Inference, and Prediction. Springer-Verlag, New York.

Hills, M. (1966). Allocation rules and their error rates. J. R. Stat. Soc. B 28, 1–31.

Hirst, D. (1996). Error-rate estimation in multiple-group linear discriminant analysis. Technometrics 38, 389–399.

Horowitz, J. L. (2000). The bootstrap. In Handbook of Econometrics (J. J. Heckman, and E. E. Leamer, eds.), Vol 5. Elsevier Science, Amsterdam, Holland.

Jain, A. K., Dubes, R. C., and Chen, C. (1987). Bootstrap techniques for error estimation. IEEE Trans. Pattern Anal. Mach. Intell. PAMI-9, 628–633.

Johnson, M. E. (1987). Multivariate Statistical Simulation. Wiley, New York.

Kanal, L. (1974). Patterns in pattern recognition: 1968–1974. IEEE Trans. Inf. Theory 2, 472–479.

Lachenbruch, P. A. (1967). An almost unbiased method of obtaining confidence intervals for the probability of misclassification in discriminant analysis. Biometrics 23, 639–645.

Lachenbruch, P. A., and Mickey, M. R. (1968). Estimation of error rates in discriminant analysis. Technometrics 10, 1–11.

Lahiri, S. N. (1994). Two term Edgeworth expansion and bootstrap approximations for multivariate studentized M-estimators. Sankhya A 56, 201–226.

Lin, M., Berg, A., and Wu, R. (2010). Modeling the genetic etiology of pharmacokinetic-pharmacodynamic links with the Arma process. J. Biopharm. Statist. 20, 351–372.

Liu, R. Y. (1988). Bootstrap procedures under some non-i.i.d. models. Ann. Stat. 16, 1696–1708.

Liu, M., Lu, W., and Tseng, C.-H. (2010). Cox regression in nested case-control studies with auxiliary covariates. Biometrics 66, 374–381.

Loh, J. M., and Jang, W. (2010). Estimating a cosmological mass bias parameter with bootstrap bandwidth selection. JRSS C Appl. Stat. 59, 761–779.

Mammen, E. (1993). Bootstrap and wild bootstrap for high dimensional linear models. Ann. Stat. 21, 255–285.

Maritz, J. S., and Jarrett, R. G. (1978). A note on estimating the variance of a sample median. J. Am. Stat. Assoc. 73, 194–196.

McLachlan, G. J. (1976). The bias of the apparent error rate in discriminant analysis. Biometrika 63, 239–244.

McLachlan, G. J. (1986). Assessing the performance of an allocation rule. Comput. Math. Appl. 12A, 261–272.

McLachlan, G. J. (1992). Discriminant Analysis and Statistical Pattern Recognition. Wiley, New York.

Miller, R. G. (1986). Beyond ANOVA, Basics of Applied Statistics. Wiley, New York.

Miller, R. G. (1997). Beyond ANOVA, Basics of Applied Statistics, 2nd Ed. Wiley, New York.

Mooney, C. Z., and Duval, R. D. (1993). Bootstrapping; a Nonparametric Approach to Statistical Inference, Vol. 95. Sage Publications, Newberry Park, CA.

Royston, P., and Altman, D. G. (1994). Regression using fractional polynomials of continuous covariates: Parsimonious parametric modeling (with discussion). Appl. Stat. 43, 429–467.

Royston, P., and Sauerbrei, W. (2003). Stability of multivariable fractional polynomial models with selection of variables and transformations: A bootstrap investigation. Stat. Med. 22, 639–659.

Royston, P., and Sauerbrei, W. (2005). Building multivariable regression models with continuous covariates in clinical epidemiology, with an emphasis on fractional polynomials. Methods Inf. Med. 44, 561–571.

Sauerbrei, W., and Royston, P. (1999). Building multivariable prognostic and diagnostic models: Transformation of the predictors by using fractional polynomials. J. R. Stat. Soc. A 162, 77–94. Corrigendum 2002; 165, 339–340.

Sauerbrei, W., and Royston, P. (2007). Modelling to extract more information from clinical trials data. Statist. Med. 26, 4989–5001.

Sen, A., and Srivastava, M. S. (1990). Regression Analysis: Theory, Methods, and Applications. Springer-Verlag, New York.

Shao, J. (1988a). On resampling methods for variance and bias estimation in linear models. Ann. Stat. 16, 986–1008.

Shao, J. (1988b). Bootstrap variance and bias estimation in linear models. Can. J. Stat. 16, 371–382.

Shao, X. (2010). The dependent wild bootstrap. J. Am. Stat. Assoc. 105, 218–235.

Shao, J., Kubler, J., and Pigeot, I. (2000). Consistency of the bootstrap procedure in individual bioequivalence. Biometrika 87, 573–585.

Shi, X. (1991). Some asymptotic results for jackknifing the sample quantile. Ann. Stat. 19, 496–503.

Shimabukuro, F. I., Lazar, S., Dyson, H. B., and Chernick, M. R. (1984). A quasi-optical experiment for measuring complex permittivity of materials. IEEE Trans. Microw. Theory Tech. 32, 659–665.

Singh, K. (1981). On the asymptotic accuracy of Efron's bootstrap. Ann. Stat. 9, 1187–1195.

Snapinn, S. M., and Knoke, J. D. (1984). Classification error rate estimators evaluated by unconditional mean square error. Technometrics 26, 371–378.

Snapinn, S. M., and Knoke, J. D. (1985a). An evaluation of smoothed classification error rate estimators. Technometrics 27, 199–206.

Snapinn, S. M., and Knoke, J. D. (1985b). Improved classification error rate estimation. Bootstrap or smooth? Unpublished report.

Srivastava, M. S., and Carter, E. M. (1983). An Introduction to Applied Multivariate Statistics. Elsevier Publishing Company, New York.

Staudte, R. G., and Sheather, S. J. (1990). Robust Estimation and Testing. Wiley, New York.

Stone, M. (1974). Cross-validatory choice and assessment of statistical predictions. J. R. Stat. Soc. B 36, 111–147.

Szechtman, H., Eckert, M. J., Tse, W. S., Boersma, J. T., Bonura, C. A., McClelland, J. Z., Culver, K. E., and Eilam, D. (2001). Compulsive checking behavior of quinpirole-sensitized rats as an animal model of obsessive-compulsive disorder (OCD): Form and control. BMC Neurosci. 2, Art.4.

Thomas, D. C. (1977). Addendum to "Methods of cohort analysis—Appraisal by application in asbestos mining" by Liddell, F. D. K., McDonald, J. C., and Thomas, D. C. J. R. Stat. Soc. A 140, 469–491.

Toussaint, G. T. (1974). Bibliography on estimation of misclassification. IEEE Trans. Inf. Theory 20, 472–479.

van de Vijver, M.J., He, Y. D., van't Veer, L. J. Dai, H., Hart, A. A. M., Voskuil, D. W., Schrieber, G. J., Peterse, J. L., Roberts, C., Marton, M. J., Parrish, M., Rodenhuis, S., Rutgers, E. T., Friend, S. H., Bernards, R.(2002). A gene-expression signature as a predictor of survival in breast cancer. N. Engl. J. Med. 347, 1999–2009.

Van Sanden, S. et al. (2011). Genomic biomarkers for a binary clinical outcome in early drug development microarray experiments. J. Biopharm. Stat., in press.

Van Sanden, S., Lin, D., and Burzykowski, T. (2007). Performance of classification methods in a microarray setting: A simulation study. Biocybern. Biomed. Eng. 27, 15–28.

van't Veer, I. J., Dai, H., van de Vijver, M. J., He, Y. D., Hart, A. A. M., Mao, M., Peterse, H. L., van der Kooy, K., Marton, M. J., Witteveen, A. T., Schreiber, G. J., Kerkhoven, R. M., Roberts, C., Linsley, P. S., Bernards, R. and Friend, S. H.(2002). Gene expression profiling predicts clinical outcomes of breast cancer. Nature 415, 530–536.

Weber, N. C. (1984). On resampling techniques for regression models. Stat. Probab. Lett. 2, 275–278.

Wu, C. F. J. (1986). Jackknife, bootstrap and other resampling plans in regression analysis (with discussion). Ann. Stat. 14, 1261–1350.

3

CONFIDENCE INTERVALS

Before we introduce the various bootstrap-type confidence intervals, we will review what confidence intervals are. In Section 3.1, we describe Efron's percentile method bootstrap and present Hartigan's typical value theorem as the motivation for Efron to introduce his version of the percentile method.

In general, what is the definition of a confidence region in two or more dimensions? Suppose we have a parameter vector v that belongs in an n-dimension Euclidean space denoted by R^n. A confidence region with confidence coefficient $1 - \alpha$ is a set of points determined on the basis of a random sample and having the property that if the random sampling were repeated infinitely many times $100(1 - \alpha)\%$ of the generated set of points representing the confidence region will contain the true parameter and $100\ \alpha\%$ will not. In the one-dimensional case, the region will be an interval or a disjoint union of intervals if it is two sided and a region of the form (a, ∞) or $(-\infty, b)$ for one-sided regions.

In parametric models where nuisance parameters (parameters required to uniquely define the probability distribution for the sample observations but that are not of interest to the investigator) are involved or in nonparametric situations, it may not be possible to construct confidence regions for v that have exactly the confidence level $1 - \alpha$ for all possible values of the nuisance parameter(s) and all possible values of v.

An Introduction to Bootstrap Methods with Applications to R, First Edition. Michael R. Chernick,
Robert A. LaBudde.
© 2011 John Wiley & Sons, Inc. Published 2011 by John Wiley & Sons, Inc.

The simplest common example of a nuisance parameter is the variance of a normal distribution when the investigator wants to estimate the mean and does not know the variance. Bahadur and Savage (1956) provided a theorem to this effect in the nonparametric setting.

None of the bootstrap confidence regions that we discuss will be exact (i.e., the actual confidence level is exactly equal to the nominal confidence level $1 - \alpha$) but they will all be consistent (as long as the estimator is), meaning that the confidence level approaches $1 - \alpha$ as the sample size gets large. If we can assume that the population distribution is symmetric, Hartigan's typical value theorem (Hartigan, 1969) demonstrates for M-estimators that exact confidence regions can be achieved using subsampling methods.

3.1 SUBSAMPLING, TYPICAL VALUE THEOREM, AND EFRON'S PERCENTILE METHOD

Consider the case of independent identically distributed observations from a symmetric distribution on the real line. Denote the n independent random variables by $X_1, X_2, \ldots,$ X_n and their distribution as F_θ. For any set A, let $P_\theta(A)$ denote the probability that a random variable X with distribution F_θ has its value in the set A. Following Efron (1982, p. 69), we assume F_θ has a symmetric density $f(\cdot)$ such that

$$P_\theta(A) = \int_A f(x - \theta)\,dx,$$

where f satisfies the conditions $\int_{-\infty}^{\infty} f(x)\,dx = 1$, $f(x) \geq 0$, and $f(-x) = f(x)$.

M-estimators are defined in Chernick (2007) and Efron (1982). For the class of M-estimators, we state Hartigan's theorem:

Theorem 1   The Typical Value Theorem. The true value of the parameter θ has probability $1/2^n$ of belonging to the interval I_i where the intervals $I_1, I_2, I_3, \ldots, I_{2^n}$ are $I_1 = (-\infty, a_1)$, $I_2 = [a_1, a_2)$, $I_3 = [a_2, a_3), \ldots, I_{2^n} = [a_{2^n-1}, \infty)$ and a_1 is the smallest M-estimate of θ out of all 2^n estimates based on the 2^n possible subsamples of the data a_2 is the second smallest, $\ldots$ and a_{2^n-1} is the largest M-estimate out of the 2^n possible subsamples.

For more details and a proof of the theorem, see Efron (1982) or Hartigan (1969). From this, Hartigan defined random subsampling as selecting at random and without replacement $B - 1$ out of the $2^n - 1$ nonempty subsets of the integers $1, 2, \ldots, n$. Partition the real line into B intervals $I_1, I_2, I_3, \ldots, I_B$ where $I_1 = (-\infty, a_1)$, $I_2 = [a_1, a_2)$, $I_3 = [a_2, a_3), \ldots, I_B = [a_{B-1}, \infty)$ and a_1 is the smallest M-estimate of θ out of all B estimates based on the B subsamples of the data a_2 is the second smallest, $\ldots$ and a_B is the largest M-estimate from the B selected subsamples.

It follows from Theorem 1 that you get an exact $100(j/B)\%$ confidence interval for θ by combining any $j \leq B - 1$ of the chosen B intervals. The most sensible way to construct a two-sided interval would be to take the j intervals in the middle. This gives an exact two-sided confidence interval.

Efron's percentile method is the most obvious way to construct a confidence interval for a parameter based on bootstrap estimates. Suppose that $\hat{\theta}_i^*$ is the ith bootstrap estimate based on the ith bootstrap sample where each bootstrap sample is of size n. By analogy to random subsampling, we would expect that if we ordered the observations from smallest to largest we would expect an interval that contains 90% of the $\hat{\theta}_i^* s$ to be a 90% confidence interval for θ.

The most sensible way to choose this confidence set would be to pick the middle 90%, taking out the lowest 5% and the highest 5%. This is precisely the interval that Efron calls a 90% percentile method confidence interval. If we used random subsampling instead of bootstrap sampling and M-estimates instead of bootstrap estimates, we would have an exact 90% confidence interval for θ based on Hartigan's result. However, this bootstrap interval is not exact because although bootstrap sampling is similar to random sampling and the M-estimates from the subsets are similar to the bootstrap estimates from the bootstrap samples, there are differences that prevent exactness, namely, bootstrapping samples with replacement from the original sample, whereas subsampling randomly selects subsets of the data and the bootstrap is applied to a wide variety of estimates, whereas Hartigan's theorem is restricted to M-estimates.

But asymptotically, the bootstrap samples behave more and more like the subsamples, and the percentile interval estimate does approach the 90% level. Unfortunately, in small to moderate samples for asymmetric or heavy-tailed distributions, the percentile method is not very good and so modifications are required to improve it. These higher-order bootstrap methods are described in Sections 3.2–3.6.

EXAMPLE 3.1   Surimi Data

We will use a common example used for the illustration of the bootstrap confidence interval estimation. "Surimi" is purified fish protein used as a material to make imitation crab and shrimp food products. The strength of surimi gels is a critical factor in production. Each incoming lot of surimi raw material is sampled and a cooked gel is prepared. From these gels, test portions are selected and tested for strength. Our sample data are the measured ultimate stresses to penetrate the test portions for 40 incoming lots of surimi. In R,

```
> set.seed(1) #for reproducible results, set random number
seed to fixed value at start
> #deformation stress required to puncture test specimens
for 40 lots of surimi
> x<- c(41.28, 45.16, 34.75, 40.76, 43.61, 39.05, 41.20,
41.02, 41.33, 40.61, 40.49, 41.77, 42.07,
+ 44.83, 29.12, 45.59, 41.95, 45.78, 42.89, 40.42, 49.31,
44.01, 34.87, 38.60, 39.63, 38.52, 38.52,
+ 43.95, 49.08, 50.52, 43.85, 40.64, 45.86, 41.25, 50.35,
45.18, 39.67, 43.89, 43.89, 42.16)
```

```
> summary(x)
Min. 1st Qu. Median Mean 3rd Qu. Max.
29.12 40.47 41.86 42.19 44.22 50.52
> c(var(x), sd(x))#variance and standard deviation of
sample
[1] 17.297605 4.159039
> shapiro.test(x)#test for normality
Shapiro-Wilk normality test
data: x
W = 0.9494, p-value = 0.07241
```

The distribution of data is roughly normal but exhibits a slight skewness to the left (note the asymmetric distances from the median to the first and third quartiles), resulting in a near significance of the Shapiro normality test (P-value $= 0.07$).

We suppose our interest here is the mean value of the ultimate stresses and the 95% confidence interval for it. Assuming normality, Student's t-based confidence interval is

```
> n<- length(x)  #sample size
> n
[1] 40
> mean(x) + qt(0.975, n-1)*sd(x)*c(-1,+1)/sqrt(n)  #student
t based 95% CI
[1] 40.85562 43.51588
```

3.2 BOOTSTRAP-*T*

In the next few sections, we will look at bootstrap improvements to Efron's percentile method that have high-order accuracy (i.e., they approach an exact confidence level at a faster rate). The bootstrap percentile-t is one of the simplest and easiest to compute, but as our colleague Kesar Singh noted (private communication), it is not used as often as BCa or ABC even though it has the same asymptotic properties.

Suppose we have a parameter θ and an estimate θ_h for it. Let θ^* be the nonparametric bootstrap estimate of θ based on the same original sample of size n that was used to obtain θ_h. Suppose that we have an estimate of the standard deviation for θ_h, say, S_h and a corresponding bootstrap estimate S^* for S_h. (Note that the catch here is that S^* must be estimable from a single bootstrap sample from which a single estimate of θ^* is found, and this estimate of S^* must be specific to the bootstrap sample. This requires some formula or special algorithm in addition to knowledge of the bootstrap sample given. Define

$$T^* = (\theta^* - \theta_h)/S^*.$$

$T*$ is a centered and standardized bootstrap version of θ (centered by subtracting θ_h and standardized by dividing by $S*$), the bootstrap analogue of

$$T = (\theta_h - \theta)/S_h.$$

If θ were the population mean and θ_h were the sample mean, the quantity T above would be our familiar t-statistic and if the sample were normally distributed, T would be a pivotal quantity. A pivotal quantity is a random quantity whose distribution is independent of the parameters in the model. In the case of T above, under the normality assumption, it has a Student's t-distribution with $n - 1$ degrees of freedom. Notice that the t-distribution depends only on the degrees of freedom, which is simply determined by known sample size n and not the population mean, θ nor the standard deviation σ (a nuisance parameter) that S_h estimates. Such quantities are called pivotal because you can make exact probability statements about them, which by a "pivoting" procedure can be converted into an exact confidence interval for θ.

To illustrate, in this case, $P\{-c < T < c\}$ is evaluated using the t-distribution with $n - 1$ degrees of freedom. Now $P\{-c < T < c\} = P\{-c < (\theta_h - \theta)/S_h < c\}$. Now the pivoting step goes as follows: $P\{-c < (\theta_h - \theta)/S_h < c\} = P\{-cS_h < (\theta_h - \theta) < cS_h\} = P\{-\theta_h - cS_h < -\theta < -\theta_h + cS_h\}$ and then multiplying all sides by -1 gives $P\{\theta_h + cS_h > \theta > \theta_h - cS_h\}$. So we have a known probability, and pivoting allows us to express it as the probability for an interval that θ belongs to.

In the case of the bootstrap statistic $T*$ for the mean under the same assumptions as above, $T*$ is asymptotical pivotal, meaning its distribution becomes independent of the parameters and its percentiles converge to the percentiles of T. This means that the same pivoting procedure leads to a confidence interval that is asymptotically accurate. As long as T and $T*$ are asymptotically pivotal, this way of generating bootstrap confidence intervals works and is more accurate than Efron's percentile method (at least for large sample sizes).

However, in the more general case when θ is a more complicated parameter than a mean, the bootstrap estimate, $\theta*$, will not be known, and so the Monte Carlo approximation is needed to generate the confidence interval. The prescription for this is as follows: for each of say B bootstrap samples, there is an estimate $\theta*$ and a $T*$ that can be calculated. Then, take the B $T*$ values and order them from smallest to largest. Apply the percentile method to T rather than to θ.

In other words, an approximate $100(1 - 2\alpha)\%$ confidence interval for theta is obtained after pivoting as $[\theta_h - T*_{(1-\alpha)}S_h, \theta_h - T*_\alpha S_h]$ where $T*_{(1-\alpha)}$ is the $100(1 - \alpha)$ percentile of the $T*$s and $T*_{(\alpha)}$ is the 100α percentile of the $T*$s. For $T*_{(1-\alpha)}$ and $T*_\alpha$ to be observed $T*$s from the bootstrap histogram of $T*$ we must have that $B\alpha$ is an integer. For arbitrary values of α, this may not be the case. When $B\alpha$ is not an integer, Efron and Tibshirani (1993) used percentiles that are neighbors. Without loss of generality, we assume $\alpha < 0.5$. So they chose the integer $k = [(B + 1)\alpha]$, the largest integer less than $(B + 1)\alpha$, and use $T*_k$ for $T*_\alpha$ and $T*_{B+1-k}$ for $T*_{(1-\alpha)}$. Now if θ_h is known to have a continuous probability density function, an exact interpolation between the integers above and below $B\alpha$ and $B(1 - \alpha)$, respectively, would work as well.

We warn the reader about terminology. There is another bootstrap confidence interval that Hesterberg et al. (2003) called "bootstrap-*t*," which only uses the bootstrap to estimate the standard error of estimate. Hesterberg recommended it only if the bootstrap distribution is approximately Gaussian with small bias. So Hesterberg's bootstrap-*t* is less general than Efron's percentile-*t*. It is defined as follows for a $100(1 - 2\alpha)\%$ confidence interval: $[\theta_h - t*S*, \; \theta_h + t*S*]$ where $t*$ is the $100(1 - \alpha)$ percentile of a Student's *t*-distribution with $n - 1$ degrees of freedom, where n is the original sample size. Notice the difference. For the percentile-*t* bootstrap, the percentiles are taken from the bootstrap distribution for $T*$ and the standard deviation is the estimate S_h from the original sample. But for Hesterberg's *t*, the percentiles are taken from Student's *t*-distribution (hence the need for normality), but the standard deviation is the bootstrap estimate of the standard deviation for θ. In Chernick and LaBudde (2010), we do simulations using Hesterberg's *t* and call it "normal-*t*" to avoid ambiguity with the bootstrap percentile-*t*.

A limitation of the bootstrap percentile-*t* is the need for S_h and the bootstrap version $S*$. When θ is a complicated parameter, we may not have an estimate at our disposal. Efron suggested that one can get around this problem by using a method called nested, iterated, or double bootstrap. We shall discuss the iterated bootstrap in the next section. The difficulty with the double bootstrap is that you are doing a bootstrap on top of a bootstrap requiring $B_1 B_2$ bootstrap iterations and assuming $B_2 > B_1$ the computational intensity is of an order higher than B_1^2. So the answer to Professor Singh's question may be that the percentile-*t* bootstrap is rarely used because for many parameters that we would do bootstrap to obtain confidence intervals, S_h is not available to us. Also double bootstrapping is rarely used because it is much more computer intensive than alternative high order accurate bootstraps.

EXAMPLE 3.2   Surimi Data Continued

For the mean, we have a formula for the standard error, namely $\sigma/\sqrt{n}$, where σ is the population standard deviation and n is the sample size. This can be estimated for any sample by $S_h = s/\sqrt{n}$. This estimate should be acceptable for moderate to large sample sizes (n at least 30 or more):

```
> set.seed(1) #for reproducibility
> u0<- mean(x) #mean of original sample
> Sh<- sd(x)/sqrt(n)
> thetas<- NULL
> tstar<- NULL
> for (i in 1:1000) { #bootstrap resampling
+ xx<- sample(x, n, replace=TRUE) #new resample
+ u<- mean(xx) #estimate
+ thetas[i]<- u #save
+ tstar[i]<- (u-u0)/(sd(xx)/sqrt(n)) #pivotal quantity
```

```
+ }
> c(u0, mean(thetas)) #compare sample mean to mean of
bootstrap sampling distribution
[1] 42.18575 42.18038
> c(Sh, sd(thetas)) #compare standard error from sample to
standard error estimate from bootstrap distribution
[1] 0.6576018 0.6262972
> summary(tstar) #look symmetric & normal?
Min. 1st Qu. Median Mean 3rd Qu. Max.
-2.90300 -0.63570 0.01188 0.01960 0.63630 3.55800
> quantile(tstar, probs=c(0.025,0.975)) #quantiles from
bootstrap percentile t
2.5% 97.5%
-1.945864 2.054761
> qt(c(0.025,0.975), n-1) #quantiles from student t
distribution
[1] -2.022691 2.022691
> u0 + quantile(tstar, probs=c(0.025,0.975))*Sh #bootstrap
percentile t confidence interval
2.5% 97.5%
40.90615 43.53696
> u0 + qt(c(0.025,0.975), n-1)*Sh #student t confidence
interval for comparison
[1] 40.85562 43.51588
```

Note that the mean of the sample (42.186) is very close to the mean of the bootstrap sampling distribution (42.180), because the mean is an unbiased estimator. The standard errors are not quite as close (0.658 vs. 0.626). The $t*$ quantiles (−1.95 and +2.05) are somewhat different from Student's t-quantiles (−2.02 and +2.02), reflecting the asymmetry of the underlying distribution. This results in a very slightly different confidence interval estimate: (40.91, 43.54) for the bootstrap and (40.86, 43.52) for Student's t-based method assuming normality. The closeness of the two intervals suggests the accuracy of both.

For comparison to the script given above, the function "boott" from the package "bootstrap" gives

```
> set.seed(1) #reproducibility
> require('bootstrap')
> boott(x, theta=mean, sdfun=function(x,nbootsd,theta,...)
{sqrt(var(x)/length(x))},
+ nboott=1000, perc=c(0.025,0.975)) #bootstrap percentile t
95% C.I.
```

```
$confpoints
0.025 0.975
[1,] 40.87091 43.44306
```

which is very close to the two intervals found above.

3.3 ITERATED BOOTSTRAP

There have been numerous contributions to the literature on bootstrap iteration. We cover their contributions in the historical notes section of the chapter. Many of the contributions come from Hall and Martin and can be found summarized up to 1992 in Hall (1992, chapter 3) and Martin (1990) as well as section 5.6 in Davison and Hinkley (1997).

Under certain regularity conditions on the distribution of the parameter estimate, Edgeworth and Cornish–Fisher expansions exist for the distribution. Based on these expansions, Hall (1988) and Martin (1990) developed an asymptotic theory for the degree of closeness of the bootstrap confidence intervals to their stated coverage probability (i.e., probability that the procedure will produce an interval that contains the true value of the parameter). An approximate one-sided confidence interval is said to be "first-order accurate" if the difference between its true coverage probability and its limiting coverage probability goes to 0 at a rate of $n^{-1/2}$. The standard interval (see Table 3.1, from page 60 of Chernick 2007) and the percentile bootstrap are both first-order accurate. An approximate one-sided confidence interval is "second-order accurate" if the difference between its true coverage probability and its limiting coverage probability goes to 0 at a rate of n^{-1}. The bootstrap percentile-t, BCa (see below), and the double bootstrap are all second-order accurate.

TABLE 3.1. Selling Price of 50 Homes Sold in Seattle in 2002

$142,000	$232,000	$132,500	$200,000	$362,000
$175,000	$50,000	$215,000	$260,000	$307,000
$197,500	$146,500	$116,700	$449,900	$266,000
$149,400	$155,000	$244,900	$66,407	$166,000
$705,000	$1,850,000	$290,000	$164,950	$375,000
$244,950	$335,000	$324,500	$222,000	$225,000
$210,950	$1,370,000	$215,500	$179,800	$217,000
$265,000	$256,000	$684,500	$257,000	$570,000
$296,000	$148,500	$270,000	$252,950	$507,000
$335,000	$987,500	$330,000	$149,950	$190,000

Given two confidence intervals that are, say, second-order accurate, which one would be the better choice? One sensible criterion would be the one that has the shortest expected length. An interval that is accurate and shortest in expected length is called "correct" according to Efron (1987). The paper also gives a very good discussion about accuracy and correctness of confidence intervals. Intervals such as BCa are both (asymptotically in sample size) second-order accurate and approximately correct.

The theory has been developed even further to show that for broad parametric families of distributions, corrections can be made to increase the order of accuracy. In fact, intervals discussed in the articles of Welch and Peers (1963), Cox and Reid (1987), and Hall (1988) are third-order accurate. It should be remembered, however, that such claims depend on knowledge of population parameters or formulas concerning them. If estimates from the sample or bootstrap samples are used in lieu of exact values, accuracy will be an asymptotic, large-sample property. Also, biases will reduce the order of accuracy for small samples. If variance estimators are needed, the sample size required to attain near asymptotic order of accuracy can be very large indeed. It should always be kept in mind that theory does not necessarily conform to practice because the assumptions may not be right or the sample size is not sufficiently large.

Bootstrap iteration provides another way to improve accuracy from second-order accuracy to third-order accuracy. Martin (1990) showed for one-sided confidence intervals each bootstrap iteration increases accuracy by one order of magnitude and by two orders of magnitude for two-sided intervals.

What is bootstrap iteration? Conceptually, it is nesting a bootstrap procedure within a bootstrap procedure, but a formal definition is not easy to understand. Suppose we have a random sample $\mathbf{X}$ consisting of observations $X_1, X_2, X_3, \ldots, X_n$. Let $X_1^*, X_2^*, X_3^*, \ldots, X_n^*$ denote a bootstrap sample from $\mathbf{X}$ and let $\mathbf{X}^*$ denote that bootstrap sample.

Let $\mathbf{I}_0$ be a nominal $1 - \alpha$ level confidence interval for a parameter φ of the population distribution from which the original random sample was taken. For example $\mathbf{I}_0$ could have been obtained by Efron's percentile method. For notational purposes to show the dependence of the original sample and the level $1 - \alpha$, we denote it as $\mathbf{I}_0(\alpha|\mathbf{X})$. Since the nominal coverage probability, $1 - \alpha$, is different from the true coverage probability (usually the true coverage is less), we denote the true coverage by $\pi_0(\alpha)$. We then let β_α be the solution to

$$\pi_0(\beta_\alpha) = P\{\varphi \in \mathbf{I}_0(\beta_\alpha|\mathbf{X})\} = 1 - \alpha. \tag{3.1}$$

Now, in Equation 3.1, replace $\mathbf{I}_0(\beta_\alpha|\mathbf{X})$ with $\mathbf{I}_0(\beta_\alpha|\mathbf{X}^*)$ and $\hat{\beta}_\alpha$ is the solution to Equation 3.1 when $\mathbf{X}^*$ replaces $\mathbf{X}$ and $\hat{\varphi}$ replaces φ. This new $\mathbf{I}_0$ is the double bootstrap interval. A triple bootstrap would involve starting with the new $\mathbf{I}_0$ in Equation 3.1 and redoing the iterative procedure just described. Although this algorithm is theoretically possible, in practice, the Monte Carlo approximation must be used and the amount of bootstrap sampling multiples as previously discussed in Section 3.2. Section 5.6 on pages 223–230 of Davison and Hinkley (1997) provides a few examples of the double bootstrap going into much detail.

EXAMPLE 3.3   Surimi Data Continued
One common use of the double bootstrap is to estimate the standard error of estimate
for each bootstrap resample when no formula is known for it. This nests the bootstrap
estimation of the standard error within the bootstrap resampling for the desired statistic.
For the Surimi data,

```
> set.seed(1) #reproducibility
> boott(x, theta=mean, nbootsd=100, nboott=1000,
perc=c(0.025,0.975)) #bootstrap percentile t 95% C.I. with
nested bootstrap
$confpoints
0.025 0.975
[1,] 40.71882 43.6630
```

Here, the standard error of each bootstrap estimate of the mean is estimated by an inner
bootstrap of 100 resamples, with 1000 resamples in the outer bootstrap. This makes the
process computation time 100 times more than before. The resulting 95% confidence
is roughly the same as before but did not require a formula for the standard error of
the mean. This method could be used therefore for any statistic, but the computation
cost is high.

3.4   BIAS-CORRECTED (BC) BOOTSTRAP

The BC bootstrap is the third in the series of four approximate confidence
interval methods shown in Table 3.1 from page 60 of Chernick (2007). The table was
a reproduction of table 6 from Efron and Tibshirani (1986). This method is designed
to work when a monotone transformation $\varphi = g(\theta)$ exist and the estimator $\hat{\varphi} = g(\hat{\theta})$
has approximately a normal distribution with mean $\varphi - z_0 \tau$. The parameter z_0 is the
bias correction and τ is the standard deviation of $\hat{\varphi}$. An estimate of z_0 is obtained by
bootstrapping.

The method for estimating z_0 by the Monte Carlo approximation to the bootstrap
is described on pages 60–61 of Chernick (2007). The BC bootstrap was shown to work
well for the bivariate correlation coefficient where the percentile method does not.
However, Schenker's (1985) example where a variance is estimated by using a sample
variance that has a chi-square distribution with 19 degrees of freedom, was a case where
the BC bootstrap does not work well, and so Efron devised BCa to get around the
problem.

3.5   BCa AND ABC

The BCa method incorporates a parameter that Efron named the "acceleration constant." It is based on the third moment and corrects for skewness based on the

Edgeworth expansion. BCa works particularly well when $\hat{\varphi}$ is approximately Gaussian with mean $\varphi - z_0 \tau_{\varphi}$, where z_0 is the bias correction just like for BC and τ_{φ} is a linear function of φ of the form $1 + a\varphi$ where a represents Efron's acceleration constant. Davison and Hinkley (1997) called BCa "the adjusted percentile method" and used different notation (w for z_0 and $\sigma(\varphi)$ for τ_{φ} with the symbol a used for the acceleration constant). However, they refer to the parameter "a" as an unknown skewness correction factor, which may be a better description of how it adjusts the interval.

The method has a second-order accuracy property that BC does not. This is true even though "a" must be estimated in the bootstrap process. Essentially, the approach is to first construct a confidence interval for φ and then using the inverse of the monotone transformation, a confidence interval for θ is generated. First, suppose that both "a" and z_0 are known.

Let $U = \varphi + (1 + a\varphi)(Z - z_0)$, where Z has a standard normal distribution with its αth percentile denoted as z_{α}. Let $U = h(T)$, the transformed estimator for $\varphi = h(\theta)$. So the random variable T is by definition $h^{-1}(U)$. The function h^{-1} exists because h is monotone. Replacing the random variables U and Z with the respective values u and z_{α}, we get

$$\hat{\varphi}_{\alpha} = u + \sigma(u)(z_{\alpha} + z_0)/[1 - a(z_{\alpha} + z_0)].$$

h is the unknown monotonic transformation mapping θ into φ. Then, $\hat{\theta}_{\alpha} = h^{-1}(\hat{\varphi}_{\alpha})$. For the bootstrap estimate T^* denote its distribution by $\hat{G}$, then

$$
\begin{aligned}
\hat{G}(\hat{\theta}_{\alpha}) &= P^*\left[T^* < \hat{\theta}_{\alpha}\big|t\right] \\
&= P^*[U^* < \hat{\varphi}_{\alpha}|u] \\
&= \Phi([\hat{\varphi}_{\alpha} - u]/\sigma(u) + z_0) \\
&= \Phi([z_0 + z_{\alpha}]/[1 - a(z_0 + z_{\alpha})] + z_0),
\end{aligned}
$$

where P^* is the probability under bootstrap sampling, and T^* and U^* are the bootstrap estimates of T and U, respectively. Since we assume "a" and z_0 are known and Φ is the cumulative standard normal distribution, the quantity on the far right-hand side of the equation is known. Therefore, a lower $100\alpha\%$ confidence bound for θ is

$$\hat{\theta}_{\alpha} = \hat{G}^{-1}\{\Phi([z_0 + z_{\alpha}]/[1 - a(z_0 + z_{\alpha})] + z_0)\}.$$

We finessed our lack of knowledge of h by using $\hat{G}$. But remember that we assumed "a" and z_0 were known. We simply replace them with bootstrap estimates. Take $\hat{z}_0 = \Phi^{-1}\{\hat{G}(t)\}$, where t is the value for which

$$P^*[T^* < t|t] = P^*[U^* < u|u] = P[U < \varphi|\varphi] = \Phi(z_0).$$

When the Monte Carlo approximation to the bootstrap is used

$$\hat{z}_0 = \Phi^{-1}(\# \text{ of } \{t_i^* \leq t\}/\{B+1\}),$$

where B is the number of bootstrap replications. From the definition of U and $\sigma(\varphi)$, we can determine that

$$\hat{a} = E*\left\{\dot{\ell}_\theta^*\left(\hat{\theta}\right)^3\right\}\Big/\left(6var*\dot{\ell}_\theta^*\left(\hat{\theta}\right)^{3/2}\right),$$

where $\dot{\ell}_\theta^*$ is the first derivative of the log likelihood l for θ where $\hat{\theta}$ is the sample estimate of θ and all expectations and likelihoods are taken under bootstrap sampling. See Davison and Hinkley (1997) for a detailed sketch of the proof that these are valid estimates of the correction factors ignoring terms of order $n-1$.

The ABC method is very close in accuracy to BCa expressing $\hat{\theta}_\alpha$ in terms of three constants denoted as a, b, and c. This expression is given on page 214 of Davison and Hinkley (1997). They derived the expression for $\hat{\theta}_\alpha$, and also for the parameters a, b, and c. Although the ABC method involves three parameters instead of the two for BCa, it is simpler and faster to calculate. As pointed out by Efron, it provides computational savings over BCa in that the constants can be estimated by numerical derivatives of log likelihood functions and from them the estimates of a and z_0 are obtained without using the Monte Carlo bootstrap replicates in the computation, saving time in the bootstrap computations.

EXAMPLE 3.4   Surimi Data Continued
Using the R package "boot,"

```
> #BCa and ABC
> set.seed(1) #for reproducibility
> require('boot')
> fboot<- function(x, i) mean(x[i]) #compute estimate given
data x and index set i
> bs<- boot(x, bootf, R=1000) #generate bootstrap estimates
> boot.ci(bs, type='bca', conf=0.95) #BCa method 95% C.I.
BOOTSTRAP CONFIDENCE INTERVAL CALCULATIONS
Based on 1000 bootstrap replicates
CALL :
boot.ci(boot.out = bs, conf = 0.95, type = "bca")
Intervals :
Level BCa
95% (40.81, 43.37 )
Calculations and Intervals on Original Scale
> fabc<- function(x, w) w%*%x #ABC uses weighted average
```

```
> abc.ci(x, fabc, conf=0.95) #ABC method C.I.
[1]  0.95000 40.84505 43.39568
```

The 95% confidence interval estimates for the BCa and ABC methods are (40.81, 43.37) and (40.85, 43.40), comparable to those found before. It should be noted that the ABC method is both very fast computationally and very accurate, typically giving intervals very close to those obtained from the BCa method.

3.6 TILTED BOOTSTRAP

Like BCa and ABC, bootstrap tilting provides second-order accurate confidence intervals. We gain accuracy over the simple percentile method (Efron's) and BC but at the cost of complication. Hesterberg et al. (2003) used real estate sale prices in the Seattle area to compare the various bootstrap confidence intervals. This is the random sample of 50 homes in Seattle that Hesterberg used.

Included in the sample frame are houses, condominiums, and commercial real estate but not plots of undeveloped land. It was chosen to illustrate a highly skewed distribution and a sample size small enough that the simple first-order methods do not work well but the second-order methods do as they adjust for the skewness better. There are three very high prices: $987,500; $1,370,000; and 1,850,000. Most of the data lie between $130,000 and $600,000 with only five values higher. Hesterberg et al. (2003) generated 1000 bootstrap samples taken from this random sample of real estate sales data. He got the following bootstrap intervals for the mean sales price:

Efron's percentile method	[252,500; 433,100]	Interval width $180,600; interval midpoint $342,800
BCa	[270,000; 455,700]	Interval width $185,700; interval midpoint $362,850
Tilting	[265,000; 458,700]	Interval width $193,700; interval midpoint $361,850

We notice that BCa and tilting provide wider intervals that are shifted to the right relative to the percentile method. We shall see that the percentile method provides less than 95% coverage because it is too narrow and centered too low. The end points for BCa and tilting are similar, but the tilting interval is wider. Their centers are close differing by only $1000, whereas the percentile method is centered approximately $20,000 lower and has the shortest width.

An advantage of tilting over BCa is that you can obtain the same level of coverage accuracy with a smaller number of bootstrap replications. Hesterberg stated that if B_1 is the number of bootstrap samples used for BCa, you can get the same accuracy with $B_2 = B_1/37$ bootstrap samples when using a tilting interval.

So what exactly is tilting? Let X_1, X_2, ..., X_n be the observed data. The original bootstrap samples with replacement from the empirical distribution function (which gives weight $1/n$ to each p_i where p_i is the probability of selecting X_i on any particular draw). Now, in bootstrap tilting, we generalize the p_i's to be arbitrary probabilities satisfying $\sum p_i = 1$ where the summation is taken over all $1 \le i \le n$ with $p_i \ge 0$ for each i. Such functions of the observations are called weighted empirical distributions.

For the tilted bootstrap, we sample with replacement from the weighted empirical distribution providing a generalization of the bootstrap. So tilting depends on the chosen multinomial distribution. One type of tilting is called maximum likelihood (ML) tilting. It picks the p_i's to solve an optimization problem, namely to maximize $\prod p_i$ subject to the constraints $\sum p_i = 1$, $p_i \ge 0$ for each i, and the parameter $\theta(p_1, p_2, \ldots, p_n) = \theta_0$ for a prespecified value θ_0. When θ is mean of the weighted empirical distribution, the solution is

$$p_i = c\left(1 - \tau\left(X_i - \theta\right)\right)^{-1},$$

where c is a norming constant that forces $\sum p_i = 1$ and τ (called the tilting parameter) is solved numerically so that $\theta(p_1, p_2, \ldots, p_n) = \theta_0$. This can be done for many other parameters. The weighted empirical distribution that is the solution to the optimization problem is called the weighted empirical distribution induced by τ and is denoted F_τ. Let U_i be the influence of X_i on θ. This is called the influence function for the parameter θ. When θ is the mean, $U_i = X_i - \theta$. An estimate for U_i is obtained by replacing θ with its sample estimate.

So in general, ML tilting gives $p_i = c(1 - \tau(U_i(\mathbf{p_0}))^{-1}$ for each i. The vector of p_i's associated with τ is denoted $\mathbf{p}_\tau$, and we write U_i as $U_i(\mathbf{p_0})$ to signify the dependence on $\mathbf{p}_\tau$ when $\tau = 0$, which is the equal weighting $\mathbf{p} = (1/n, 1/n, \ldots, 1/n)$. In addition to ML tilting, another method is called exponential tilting, which corresponds to a different optimization equation. For exponential tilting, the solution is

$$p_i = c\exp\left(\tau U_i\left(\mathbf{p_0}\right)\right).$$

After generating bootstrap samples from F_τ, the ordinary percentile method is applied to the bootstrap samples from this weighted empirical distribution function. In order to keep things simple in his companion chapter, Hesterberg et al. (2003), did not discuss the tilted bootstrap in any detail and did not mention the various families of distributions that result from different optimization problems.

So it is not clear without replicating the process for exponential and ML tilting, which tilting interval he generated. The method of sampling is called importance sampling because the higher weights are presumably given to the part of the distribution that is most important for obtaining an accurate estimate. Usually, importance sampling gives higher weight to the tails of the distribution so that the extreme part of the distribution is better represented in the Monte Carlo samples. Use of importance sampling can reduce the variance of the estimate and hence for a given accuracy you can obtain the results with fewer bootstrap samples.

3.7   VARIANCE ESTIMATION WITH SMALL SAMPLE SIZES

Robust estimators of variance have long been a challenge to researchers. Levene's method, which goes back to the 1960s, is still one of the best, and it has had numerous modifications. Because bootstrap confidence intervals have been successful at providing approximately the desired coverage probability for various parameters over a wide range of population distributions, we might be tempted to think that they would work well for standard deviations and variances also, especially BCa, ABC, double bootstrap, percentile-t, and the tilted bootstrap, which all have the second-order accuracy property. However, second-order accuracy is an asymptotic property.

Can we expect the same performance in small samples? Some researchers have tried the bootstrap with skewed and heavy-tailed distributions and found that the coverage was surprisingly poor in small to moderate sample sizes. In an interesting paper, Schenker (1985) showed that for a particular chi-square distribution, associated with the variance parameter, the bootstrap percentile method and Efron's BC did not perform well. BCa and other second-order accurate bootstrap confidence intervals were constructed to get around this problem.

Given the recent problems encountered with even the second-order methods, we were motivated to do extensive simulations of several bootstrap confidence intervals for various sample sizes to better understand the issue. In essence, we were revisiting the problem raised by Schenker (1985). In Efron and Tibshirani (1986), a table is given showing the progression of nested bootstrap methods that apply more broadly as more parameters are introduced to correct the end points and obtain higher-order accuracy.

Our simulations show for asymmetric and heavy-tailed families of distribution the coverage probability even for the high-order bootstraps can be grossly overestimated. Our results appeared in Chernick and LaBudde (2010). Another surprising result is that the lower-order bootstraps that are less complicated and involve estimation of fewer parameters sometimes give better coverage probability even though they do not converge at the fastest rate.

Our investigation through simulation confirms the asymptotic results but sometimes shows a different order of preference in the small sample size setting ($10 \le n \le 50$). Simulation studies addressing the small sample properties of specific bootstrap estimates occur sporadically in the literature. The two books Shao and Tu (1995) and Manly (2006) are two examples where such simulations are included. R code for lognormal distribution is presented in the appendix of Chernick and LaBudde (2010).

The key parameters of the simulations are as follows:

nSize. The sample size of the originating data that is to be bootstrapped.

nReal. The number of bootstrap resamples used to estimate the bootstrap resampling distribution (the number of possible unique resamples is always no more than $nSize^{nSize}$).

nRepl. The number of Monte Carlo replications of the entire experiment, based on generating new samples of size *nSize* from the underlying assumed distribution, in order to estimate coverage accuracy and other errors of the bootstrap methodology.

In the article, we reported on the following bootstrap methods:

Normal-t. A parametric Student's t confidence interval, with center point the sample variance and the standard error of variance estimated from that of the resampling distribution. This differs from a parametric normal bootstrap in that the percentile of the t-distribution with $n - 1$ degrees of freedom is used instead of the standard normal percentile. In Hesterberg's bootstrap chapter (Hesterberg et al., 2003), it is referred to as the bootstrap-t, but that confuses it with the bootstrap percentile t presented earlier in the chapter.

EP. Efron percentile interval, with end points the plug-in quantiles of the resampling distribution.

BC. The Efron BC interval. Simulations have shown that the BC method is, as expected, virtually identical in estimates to the BCa interval with the acceleration $a = 0$ (i.e., adjustment for median bias only).

BCa. The Efron bias-corrected-and-accelerated interval, with median bias correction and skew correction via a jackknife estimate of the (biased) coefficient of skewness from the original sample.

ABC. The Efron–DiCiccio approximate bootstrap confidence interval.

We compared bootstrap estimates for Gamma(2, 3), Uniform(0, 1), Student's t with 5 degrees of freedom, Normal(0, 1) and Lognormal(0, 1) for various coverage probabilities. For a thorough account, see Chernick and LaBudde (2010). Here, we will focus on the worst case the lognormal and the best case the uniform. See Tables 3.2–3.4.

For the Uniform(0, 1) distribution we simulate two cases:

1. The sample size *nSize* is 25, and 1000 bootstrap samples (*nReal*) of *nSize* 25 were generated by Monte Carlo. We considered 1000 Monte Carlo replications and also 64,000 replications (*nRepl*) for each case. This provides an example of a short-tailed distribution. In this case, we compare the normal-t with Efron's percentile method (Table 3.2).

2. The sample size is varied from 10 to 100 and we compare normal-t, Efron's percentile method, ABC, and BCa with sample size varied. For both (1) and (2), we used confidence levels of 50%, 60%, 70%, 80%, 90%, 95%, and 99%. We only show 90%. See Chernick and LaBudde (2010) for more details.

These tables show for the specified nominal coverage probabilities the Monte Carlo estimates of the actual coverage probabilities for the various intervals.

We observe that EP and normal-t were the best in general for this distribution, so in (1), we only compared these two and found that the coverage is most accurate at 50% and 60%. For 80% and up, the actual coverage is considerably below the nominal coverage, which is due to the small sample size of 25. In (2), we see that for a confidence level of 90%, the sample size needs to be 50 or higher for the accuracy to be good.

TABLE 3.2. Uniform(0, 1) Distribution: Results for Various Confidence Intervals

(1) Results: nSize = 25, nReal = 1000		
Confidence level (%)	Normal-*t* (%)	EP (%)
50	49.5	49
60	60.4	57.3
70	67.2	68
80	78.4	77.9
90	87.6	86.8
95	92.7	92
99	97.4	96.5

Source: Taken from table 4 in Chernick and LaBudde (2010).

TABLE 3.3. ln[Normal(0, 1)] Distribution: Results for Nominal Coverage Ranging 50

Nominal confidence level (%)	*nRepl*	*nReal*	Normal-*t* (%)	EP (%)	BC (%)	ABC (%)	BCa (%)
50	64,000	16,000	24.91	32.42	21.86	24.34	21.99
60	64,000	16,000	31.45	35.87	35.32	30.08	35.95
70	64,000	16,000	38.91	38.91	41.49	36.38	43.07
80	64,000	16,000	44.84	43.74	46.70	43.90	48.71
90	64,000	16,000	50.32	50.11	52.52	53.03	56.43
95	64,000	16,000	53.83	53.06	56.60	59.09	61.66
99	64,000	16,000	60.05	59.00	61.68	65.43	67.29

Again, we see that EP and normal-*t* are slightly better than the others especially when *n* is 20 or less. Also, in almost every case for each estimation procedure, we find that the true coverage is less than the nominal coverage. There are a few exceptions where the coverage is slightly higher than nominal.

Now, we present the results for the lognormal. In this example from Table 3.4, we considered *nSize* = 25. Sixty-four thousand (*nRepl*) samples of size 25 were generated by Monte Carlo. We considered 16,000 bootstrap iterations for each case and compare normal-*t*, Efron's percentile method, BC, ABC, and BCa. We used confidence levels of 50%, 60%, 70%, 80%, 90%, 95%, and 99% for the comparisons. This is an example of a highly skewed distribution for the original data.

Table 3.4 shows the results for the lognormal when the sample size is varied from 10 to 2000 for nominal coverage ranging from 50% to 99%.

From Table 3.4, we see that all methods overestimate coverage (nominal coverage higher than actual coverage) by a factor of 2 when *nSize* = 10. For *nSize* = 100, these methods overestimate coverage by 10% or more. Even for *nSize* = 2000, coverage error is still 3.5% or more for the best method (BCa at 99% nominal confidence). For nominal confidence levels from 50% to 60%, the EP method generally performs best. For confidence levels of 60% or more, the BCa method is generally best in accuracy (at 60%, EP and BCa are virtually equal).

TABLE 3.4. ln[Normal(0,1)] Distribution

nSize	nRepl	nReal	Normal-t (%)	EP (%)	BC (%)	ABC (%)	BCa (%)
			Results for 50% nominal coverage				
10	64,000	16,000	17.86	25.26	20.11	16.88	18.37
25	64,000	16,000	24.91	32.42	21.86	24.34	21.99
50	16,000	16,000	29.69	35.22	26.49	28.98	27.01
100	16,000	16,000	33.38	37.74	32.27	32.74	31.60
250	16,000	16,000	36.56	39.60	36.36	36.29	35.58
500	16,000	16,000	39.65	42.03	39.09	38.46	38.81
1000	16,000	16,000	41.33	43.18	41.06	*	40.72
2000	16,000	16,000	43.23	44.69	42.48	*	42.46
			Results for 60% coverage				
10	64,000	16,000	22.68	28.61	26.52	20.84	25.78
25	64,000	16,000	31.45	35.87	35.32	30.08	35.95
50	16,000	16,000	37.02	40.28	39.38	35.59	40.24
100	16,000	16,000	41.51	43.76	43.19	40.13	43.55
250	16,000	16,000	45.21	46.80	46.42	44.42	46.68
500	16,000	16,000	48.37	49.76	48.98	49.94	49.28
1000	16,000	16,000	50.74	51.59	50.99	*	51.07
2000	16,000	16,000	52.85	53.64	52.24	*	52.13
			Results for 70% coverage				
10	64,000	16,000	28.64	30.59	31.30	25.32	32.54
25	64,000	16,000	38.91	39.07	41.49	36.38	43.07
50	16,000	16,000	44.89	44.54	46.62	42.96	48.11
100	16,000	16,000	49.74	49.49	50.75	47.98	52.49
250	16,000	16,000	54.21	54.10	54.74	53.12	55.69
500	16,000	16,000	57.83	57.68	57.94	55.50	58.63
1000	16,000	16,000	60.36	60.37	59.97	*	60.37
2000	16,000	16,000	62.54	62.38	61.47	*	61.67
			Results for 80% coverage				
10	64,000	16,000	35.04	33.76	35.01	30.51	36.74
25	64,000	16,000	44.84	43.74	46.70	43.90	48.71
50	16,000	16,000	51.14	50.51	53.11	51.49	55.34
100	16,000	16,000	56.48	56.19	58.61	57.38	60.54
250	16,000	16,000	62.26	62.14	63.29	63.06	64.81
500	16,000	16,000	66.24	66.08	67.10	65.70	68.11
1000	16,000	16,000	69.31	69.03	69.35	*	69.80
2000	16,000	16,000	71.80	71.40	71.28	*	71.58
			Results for 90% coverage				
10	64,000	16,000	39.98	37.12	39.38	37.11	41.03
25	64,000	16,000	50.32	50.11	52.52	53.03	56.13

(Continued)

TABLE 3.4. (Continued)

nSize	nRepl	nReal	Normal-t (%)	EP (%)	BC (%)	ABC (%)	BCa (%)
50	16,000	16,000	56.93	57.39	60.43	62.04	64.63
100	16,000	16,000	62.93	63.93	66.71	68.50	70.27
250	16,000	16,000	69.35	70.56	72.74	74.41	75.33
500	16,000	16,000	74.13	74.99	76.63	77.09	78.69
1000	16,000	16,000	77.63	78.40	79.59	*	80.81
2000	16,000	16,000	80.38	80.85	81.83	*	82.36
			Results for 95% coverage				
10	64,000	16,000	43.36	39.53	41.56	39.21	44.58
25	64,000	16,000	53.83	53.06	56.60	59.09	61.66
50	16,000	16,000	60.94	61.66	65.26	69.02	70.65
100	16,000	16,000	67.13	68.88	71.26	75.70	76.82
250	16,000	16,000	73.96	75.94	78.18	81.83	82.01
500	16,000	16,000	78.59	80.56	82.58	84.07	85.26
1000	16,000	16,000	82.23	83.72	85.23	*	86.97
2000	16,000	16,000	85.06	86.48	87.43	*	88.85
			Results for 99% coverage				
10	64,000	16,000	49.51	42.84	44.92	44.14	47.83
25	64,000	16,000	60.05	59.00	61.68	65.43	67.29
50	16,000	16,000	67.64	68.90	71.58	77.44	78.06
100	16,000	16,000	74.11	76.45	78.99	84.69	84.82
250	16,000	16,000	80.93	83.71	85.47	90.16	89.82
500	16,000	16,000	85.53	88.16	89.41	92.24	92.68
1000	16,000	16,000	88.49	90.74	91.77	*	94.20
2000	16,000	16,000	91.31	93.13	93.83	*	95.49

*Omitted calculations due to excessive computational time.
Source: Taken from Chernick and LaBudde (2010), table 8.

As *nSize* becomes large, the differences among methods shrink in size. For *nSize* small (10 or less), the normal-*t* method performs best, probably because *nSize* is too small for the generation of reasonable resampling distributions. Regardless of these comments, it should be noted that all methods have large coverage errors for *nSize* = 100 or less, and this does not improve much even for *nSize* as large as 2000.

The fact that the results for the lognormal distribution were far worse in undercoverage than any of the others examples we considered is a little bit of a surprise. The heavy skewness has something to do with it. But more research may be required to understand this.

3.8 HISTORICAL NOTES

Bootstrap confidence intervals were introduced in Efron (1979, 1981, 1982, chapter10). The bootstrap percentile-*t* method was first discussed in Efron (1982), and the methods were illustrated using the median as the parameter of interest. The BCa interval was first given in Efron (1987) in response to the example in Schenker (1985). For the BC interval estimates, Buckland (1983, 1984, 1985) provided applications along with algorithms for their construction.

At that time, Efron and others recognized that confidence interval estimation was more difficult than bootstrap estimates of standard errors, and more bootstrap samples were recommended to be generated for confidence intervals. Recommendations at that time were to have at least 100 bootstrap samples for standard errors but at least 1000 for confidence intervals.

Schenker (1985) was the first paper to send a cautionary note about confidence intervals and showed that BC was not sufficiently accurate. This led to the BCa and other second-order accurate intervals that appeared in Efron (1987).

The idea of iterating the bootstrap to get higher-order accuracy appears in Hall (1986), Beran (1987), Loh (1987), Hall and Martin (1988), and DiCiccio and Romano (1988). The methods in these papers differ in the way they adjust the interval end point(s). Loh introduced the name bootstrap calibration for his procedure.

The issue of sample size requirements is covered in Hall (1985). Specific application of these confidence interval methods to the Pearson correlation coefficient is given in Hall et al. (1989). Further developments of bootstrap iteration can be found in Martin (1990), Hall (1992), and Davison and Hinkley (1997).

Much of the asymptotic theory for bootstrap confidence intervals is based on formal Edgeworth expansions. Hall (1992) provided a detailed account of the application of these techniques to bootstrap confidence intervals where he found them to be very useful in obtaining the order of accuracy for the intervals.

Saddlepoint approximations provide another way to look at the theory of confidence intervals providing accurate intervals competitive with the high-order bootstrap methods but not requiring the bootstrap Monte Carlo replicates. These ideas are covered in Field and Ronchetti (1990), Davison and Hinkley (1988), and Tingley and Field (1990). Other methods that provide confidence intervals without the need for the Monte Carlo approximation to the bootstrap distribution for the estimator are DiCiccio and Efron (1992) and DiCiccio and Romano (1989).

Techniques to reduce the number of bootstrap replications (variance reduction methods) are provided in a later chapter in this text as well as the articles by Davison et al. (1986), Therneau (1983), Hesterberg (1988), Johns (1988), and Hinkley and Shi (1989). Importance resampling is one of the key methods described in these papers and that idea is exploited in the tilted bootstrap method that we encountered earlier in this chapter. The most recent developments can be found in Hesterberg (1995a, b, 1996, 1997) and Hesterberg et al. (2003) for an introduction to the tilted bootstrap.

Asymptotically pivotal quantities produce high-order accuracy for bootstrap confidence intervals when they can be found. Their application to bootstrap confidence intervals was first mentioned in a series of papers (Babu and Singh, 1983, 1984, 1985). Hall (1986) and Hall (1988) both demonstrated the increased accuracy for bootstrap confidence intervals that use asymptotically pivotal quantities over ones that do not.

Hall (1986) provided the asymptotic rate of decrease in coverage error as the sample size gets large for the bootstrap percentile-t method (an example where an asymptotically pivotal quantity is used). Beran (1987) and Liu and Singh (1987) also provided a theory to support the use of asymptotically pivotal quantities when they are available. Much of Hall's work is also given in his book (Hall, 1992).

Difficulties with bootstrap confidence intervals for variances were identified in the early 2000s. This work motivated us to provide a detailed simulated account of the undercoverage in Chernick and LaBudde (2010).

3.9   EXERCISES

1. Consider the mice weight data of Chapter 1 (Exercise 1.6.1):
   (a) Find a 95% confidence interval on the mean using the standard Student's t-distribution.
   (b) Find a 95% confidence interval on the mean using Efron's percentile method.
   (c) Find a 95% confidence interval on the mean using the BCa method (and the ABC method, if you have a means of doing so).
   (d) Find a 95% confidence interval on the mean using the percentile-t method.
   (e) How do the intervals compare? Why are the intervals for Exercise 3.9.1a,d so much larger than those for Exercise 3.9.1b,c? What does this tell you about the benefits of parametric methods on small ($n < 30$) samples and the problems of using bootstrap on such samples? What does it tell you about the percentile-t method compared with the other bootstrap methods?

2. Consider the mice maze transit time data of Chapter 1 (Exercise 1.6.2):
   (a) Find a 95% confidence interval on the mean using the standard Student's t-distribution.
   (b) Find a 95% confidence interval on the mean using Efron's percentile method.
   (c) Find a 95% confidence interval on the mean using the BCa method (and the ABC method, if you have a means of doing so).
   (d) Find a 95% confidence interval on the mean using the percentile-t method.
   (e) How do the intervals compare? Why are the intervals for Exercise3.9.2a,d so much larger than those for Exercise3.9.2b,c? What does this tell you about the benefits of parametric methods on small ($n < 30$) samples and the problems of using bootstrap on such samples? What does it tell you about the percentile-t method compared with the other bootstrap methods?

3. Consider the aflatoxin residues in peanut butter data of Chapter 1 (Exercise 1.6.5):
   (a) Find a 95% confidence interval on the mean using the standard Student's t-distribution.
   (b) Find a 95% confidence interval on the mean using Efron's percentile method.
   (c) Find a 95% confidence interval on the mean using the BCa method (and the ABC method, if you have a means of doing so).
   (d) Find a 95% confidence interval on the mean using the percentile-t method.

(e) How do the intervals compare? Which intervals do you trust? What does this tell you about the benefits of parametric methods on small ($n < 30$) samples and the problems of using bootstrap on such samples? What does it tell you about the percentile-t method compared with the other bootstrap methods, at least when a formula for the standard error is known?

4. Consider the razor blade sharpness data of Chapter 1 (Exercise 1.6.6):
   (a) Find a 95% confidence interval on the mean using the standard Student's t-distribution.
   (b) Find a 95% confidence interval on the mean using Efron's percentile method.
   (c) Find a 95% confidence interval on the mean using the BCa method (and the ABC method, if you have a means of doing so).
   (d) Find a 95% confidence interval on the mean using the percentile-t method.
   (e) How do the intervals compare? Which intervals do you trust? What does this tell you about the benefits of parametric methods on small ($n < 30$) samples and the problems of using bootstrap on such samples? What does it tell you about the percentile-t method compared with the other bootstrap methods, at least when a formula for the standard error is known?

5. Consider the airline accident data of Chapter 2 (Exercise 2.8.1):
   (a) Find a 95% confidence interval on the mean using the standard Student's t-distribution.
   (b) Find a 95% confidence interval on the mean using Efron's percentile method.
   (c) Find a 95% confidence interval on the mean using the BCa method (and the ABC method, if you have a means of doing so).
   (d) Find a 95% confidence interval on the mean using the percentile-t method.
   (e) How do the intervals compare? Which intervals do you trust? What does this tell you about the benefits of parametric methods on small ($n < 30$) samples and the problems of using bootstrap on such samples? What does it tell you about the percentile-t method compared with the other bootstrap methods, at least when a formula for the standard error is known?

6. Microbiological method comparison: A laboratory wants to determine if two different methods give similar results for quantifying a particular bacterial species in a particular medium. A quick test is performed by a decimal serial dilution assay, using each method in replicate on each dilution. The dilution with the highest usable counts (falling the range 30–300) had results (actual data):

Method	Replicate							
	1	2	3	4	5	6	7	8
A	176	125	152	180	159	168	160	151
B	164	121	137	169	144	145	156	139

Such counts typically are modeled using either a Poisson distribution or, more commonly, a lognormal distribution. Neither of these are exactly correct for the application, but both fit well in practice.

(a) Transform the data using $Y = \log10(\text{Count} + 0.1)$. (The 0.1 protects against zero count data that sometimes occurs.)

(b) For the original counts and the log10-transformed counts, find a 95% confidence interval in the usual way on the mean difference, in each case assuming the data are normally distributed. Assume the replicates are unpaired.

(c) For the original counts and the log10-transformed counts, find a 95% confidence intervals by bootstrapping with 1000 resamples, using Efron's percentile method. (Hint: Resample A and B separately, and then combine.)

(d) Compare the confidence intervals found in Exercise 3.9.6b,c. Interpret the effect of the transform and the difference between the parametric and bootstrap approaches.

REFERENCES

Babu, G. J., and Singh, K. (1983). Nonparametric inference on means using the bootstrap. Ann. Stat. 11, 999–1003.

Babu, G. J., and Singh, K. (1984). On one term Edgeworth correction by Efron's bootstrap. Sankhya A 46, 195–206.

Babu, G. J., and Singh, K. (1985). Edgeworth expansions for sampling without replacement from finite populations. J. Multivar. Anal. 17, 261–278.

Bahadur, R., and Savage, L. (1956). The nonexistence of certain statistical procedures in nonparametric problems. Ann. Math. Stat. 27, 1115–1122.

Beran, R. J. (1987). Prepivoting to reduce level error of confidence sets. Biometrika 74, 457–468.

Buckland, S. T. (1983). Monte Carlo methods for confidence interval estimation using the bootstrap technique. BIAS 10, 194–212.

Buckland, S. T. (1984). Monte Carlo confidence intervals. Biometrics 40, 811–817.

Buckland, S. T. (1985). Calculation of Monte Carlo confidence intervals. Appl. Statist. 34, 296–301.

Chernick, M. R. (2007). Bootstrap Methods: A Guide for Practitioners and Researchers, 2nd Ed. Wiley, Hoboken, NJ.

Chernick, M. R., and LaBudde, R. A. (2010). Revisiting qualms about bootstrap confidence intervals. Am. J. Math. Manag. Sci. 29, 437–456.

Cox, D. R., and Reid, N. (1987). Parameter orthogonality and approximate conditional inference (with discussion). J. R. Stat. Soc. B 49, 1–39.

Davison, A. C., and Hinkley, D. V. (1988). Saddlepoint approximations in resampling methods. Biometrika 75, 417–431.

Davison, A. C., and Hinkley, D. V. (1997). Bootstrap Methods and Their Applications. Cambridge University Press, Cambridge.

Davison, A. C., Hinkley, D. V., and Schechtman, E. (1986). Efficient bootstrap simulation. Biometrika 73, 555–566.

DiCiccio, T. J., and Efron, B. (1992). More accurate confidence intervals in exponential families. Biometrika 79, 231–245.

DiCiccio, T. J., and Romano, J. P. (1988). A review of bootstrap confidence intervals (with discussion). J. R. Stat. Soc. B 50, 338–370. Correction J. R. Stat. Soc. B 51, 470.

DiCiccio, T. J., and Romano, J. P. (1989). The automatic percentile method: Accurate confidence limits in parametric models. Can. J. Stat. 17, 155–169.

Efron, B. (1979). Bootstrap methods: Another look at the jackknife. Ann. Stat. 7, 1–26.

Efron, B. (1981). Nonparametric standard errors and confidence intervals (with discussion). Can. J. Stat. 9, 139–172.

Efron, B. (1982). The Jackknife, the Bootstrap and Other Resampling Plans. SIAM, Philadelphia.

Efron, B. (1987). Better bootstrap confidence intervals (with discussion). J. Am. Stat. Assoc. 82, 171–200.

Efron, B., and Tibshirani, R. (1986). Bootstrap methods for standard errors, confidence intervals and other measures of statistical accuracy. Stat. Sci. 1, 54–77.

Efron, B., and Tibshirani, R. (1993). An Introduction to the Bootstrap. Chapman and Hall, New York.

Efron, B. (1990). More efficient bootstrap computations. J. Am. Statist. Assoc. 85, 45–58.

Field, C., and Ronchetti, F. (1990). Small Sample Asymptotics. Institute of Mathematical Statistics, Hayward, CA.

Hall, P. (1985). Recovering a coverage pattern. Stoch. Proc. Appl. 20, 231–246.

Hall, P. (1986). On the bootstrap and confidence intervals. Ann. Stat. 14, 1431–1452.

Hall, P. (1988). Theoretical comparison of bootstrap confidence intervals (with discussion). Ann. Stat. 16, 927–985.

Hall, P. (1992). The Bootstrap and Edgeworth Expansion. Springer-Verlag, New York.

Hall, P., and Martin, M. A. (1988). On bootstrap resampling and iteration. Biometrika 75, 661–671.

Hall, P., Martin, M. A., and Schucany, W. R. (1989). Better nonparametric bootstrap confidence intervals for the correlation coefficient. J. Stat. Comput. Simul. 33, 161–172.

Hartigan, J. A. (1969). Using subsample values as typical values. J. Am. Stat. Assoc. 64, 1303–1317.

Hesterberg, T. (1988). Variance reduction techniques for bootstrap and other Monte Carlo simulations. PhD dissertation, Department of Statistics, Stanford University.

Hesterberg, T. (1995a). Tail-specific linear approximations for efficient simulations. J. Comput. Graph. Stat. 4, 113–133.

Hesterberg, T. (1995b). Weighted average importance samplingand defense mixture distributions. Technometrics 37, 185–194.

Hesterberg, T. (1996). Control variates and importance sampling for efficient bootstrap simulations. Stat. Comput. 6, 147–157.

Hesterberg, T. (1997). Fast bootstrapping by combining importance sampling and concomitants. Proc. Comput. Sci. Stat. 29, 72–78.

Hesterberg, T., Monaghan, S., Moore, D. S., Clipson, A., and Epstein, R. (2003). Bootstrap Methods and Permutations Tests: Companion Chapter 18 to the Practice of Business Statistics. W. H. Freeman and Company, New York.

Hinkley, D. V., and Shi, S. (1989). Importance sampling and the bootstrap. Biometrika 76, 435–446.

Johns, M. V. (1988). Importance sampling for bootstrap confidence intervals. J. Am. Stat. Assoc. 83, 709–714.

Liu, R. Y., and Singh, K. (1987). On partial correction by the bootstrap. Ann. Stat. 15, 1713–1718.

Loh, W.-Y. (1987). Calibrating confidence coefficients. J. Am. Stat. Assoc. 82, 155–162.

Manly, B. F. J. (2006). Randomization, Bootstrap and Monte Carlo Methods in Biology, 3rd Ed. Chapman & Hall/CRC, Boca Raton, FL.

Martin, M. A. (1990). On the bootstrap iteration for coverage correction in confidence intervals. J. Am. Stat. Assoc. 85, 1105–1118.

Schenker, N. (1985). Qualms about bootstrap confidence intervals. J. Am. Stat. Assoc. 80, 360–361.

Shao, J., and Tu, D. (1995). The Jackknife and Bootstrap. Springer-Verlag, New York.

Therneau, T. (1983). Variance reduction techniques for the bootstrap. PhD dissertation, Department of Statistics, Stanford University.

Tingley, M., and Field, C. (1990). Small-sample confidence intervals. Am. Stat. Assoc. 85, 427–434.

Welch, W. J., and Peers, H. W. (1963). On formulae for confidence points based on integrals of weighted likelihoods. J. R. Stat. Soc. B 25, 318–329.

4

HYPOTHESIS TESTING

There are many ways to do hypothesis testing by the bootstrap. Perhaps the simplest is the inversion of a bootstrap confidence interval, which takes advantage of the 1-1 correspondence of confidence intervals and hypothesis tests. This will be discussed in Section 4.1.

We briefly review the basic ideas of the Neyman–Pearson approach to hypothesis testing. Usually, in research, we are interested in an outcome variable (in clinical trials, this is called an end point). A null hypothesis is formed with the hope of rejecting it. Often, the null hypothesis states that the endpoint parameter is zero.

For example, in a clinical trial, the end point could be the difference in response to a drug, where we are interested in the difference between the mean for the population of patients getting the drug and the population of patients on placebo. A value of zero is uninteresting since it indicates that the drug is ineffective.

It may be that a small difference is also not good as the drug may not be worth marketing if it provides little improvement over placebo. The magnitude of the difference that is large enough to market the drug is often referred to in the pharmaceutical industry as the clinically significant difference.

In general, hypothesis tests can be one sided or two sided. For a two-sided test, we are interested in knowing that there is a large difference regardless of whether the

An Introduction to Bootstrap Methods with Applications to R, First Edition. Michael R. Chernick, Robert A. LaBudde.
© 2011 John Wiley & Sons, Inc. Published 2011 by John Wiley & Sons, Inc.

difference is positive or negative. This would be the case if the two treatments are competing drugs and we want to know which one is more effective.

For a one-sided test, we are only interested in one direction. So in the case where we are comparing a new drug to placebo, it is not interesting to us if placebo is more effective than the drug. If for high values of the end point mean the treatment is more effective than for low values and we compute drug effect—placebo effect, then only positive differences would be of interest and we would want to do a one-sided test.

To apply hypothesis testing to our research question, we compute a test statistic. In the case of the drug versus placebo parallel group trial that we have been describing, the test statistic is simply the appropriately normalized difference of the sample means for each group. Often, the subject in a parallel design is placed in a particular group at random. Usually, the randomization is 1:1, meaning the sample size is the same for both groups. But in the Tendril DX example in Section 4.2, we shall see that sometimes there can be practical reasons for doing 2:1 or even 3:1 randomization of treatment to control (e.g., drug to placebo or new device to previous device).

Once the test statistic is determined, we need to identify its sampling distribution when the null hypothesis is assumed to be correct (i.e., how the test statistic would vary in repeated independent experiments when the null hypothesis is correct). To obtain the sampling distribution, we must have normalized the test statistic to eliminate nuisance parameters (e.g., the variance of the mean difference). Then, the sampling distribution is determined by

1. making a parametric assumption,
2. applying asymptotic theory (e.g., the central limit theory),
3. bootstrapping the distribution "under the null hypothesis."

You can find examples of 1 and 2 in any basic statistic text that covers hypothesis testing. Here of course we will concentrate on 3. We know how to bootstrap to get the actual bootstrap sampling distribution, but in hypothesis testing, the introduction of the null hypothesis raises some issues.

To complete the description we need to define the type I and type II error rates. The type I error rate is the probability of falsely rejecting the null hypothesis. In the Neyman–Pearson approach, you fix the type I error by defining critical values for the test statistic. For one-sided tests, there will be one critical value, and for two-sided tests, there will be two (usually C and $-C$, but in general, C_1 and C_2). The relationship between the two critical values will depend on whether the sampling distribution under the null hypothesis is symmetric around zero or not. The type I error is a probability and is denoted as α. We call the type I error the level of the test. It is common practice but not mandatory to take $\alpha = 0.05$.

The type II error is the probability of not rejecting the null hypothesis when the null hypothesis is not correct. Often, the null hypothesis constitutes a single point in the parameter space. In the case of the parallel design, the value is 0. Now, as stated, the type II error is not well defined. We must first pick a point in the parameter space different from 0, and then determine the sampling distribution when this alternative is true.

The portion of the distribution that is in the region where the null hypothesis is accepted (we will explain and illustrate this later) is then the type II error. Now, the sampling distribution depends on the value of the parameter that we picked, so there is actually a different type II error for each value. If the value we pick is very close to zero, the type II error will be close to $1 - \alpha$. As we get very far away from zero, it will approach 0.

The type II error is denoted by β. We call the probability of correctly rejecting the null hypothesis when an alternative value is correct (a value different from 0 in our parallel treatment design) the power of the test (under the particular alternative value). The power is $1 - \beta$. When we are close to zero in our example, $\beta \approx 1 - \alpha$, and so $1 - \beta \approx \alpha$. For sensible hypothesis tests, the power will always be greater than α and increase monotonically to 1 as the parameter moves away from 0 in one direction.

Another term commonly used in hypothesis testing is the P-value. The P-value is the probability of observing a more extreme value for the test statistic than the one actually observed when the null hypothesis is true. If your test statistic is exactly equal to the critical point, the P-value equals α. Whenever we reject the null hypothesis, the P-value is less than or equal to α. If we fail to reject, the P-value will be bigger than α.

The power of the test at any particular alternative also depends on the sample size since the variance gets smaller as the sample size increases. The power is usually chosen to be high, 0.80, 0.90, or 0.95, at a clinically meaningful difference between the two groups. The critical values are chosen to achieve the specified type I error rate regardless of what sample size we use. Then, to achieve 0.80 power when the true difference is 1, we must have a minimum sample size N (1). There are computer programs and tables that can determine this minimum sample size for you when you specify all the above parameters.

4.1 RELATIONSHIP TO CONFIDENCE INTERVALS

The 1-1 relationship between confidence intervals and hypothesis tests make it possible to construct a bootstrap test using any bootstrap method to obtain a confidence interval. However, the significance level of the test is related to the confidence level, and so you would want to pick an accurate bootstrap confidence interval so that the level of the test will be approximately what it is intended to be.

The correspondence is as follows: Suppose you have a $100(1 - \alpha)\%$ bootstrap confidence interval; then, if the null hypothesis H_0 is $\theta = \theta_0$, you obtain a bootstrap test by rejecting H_0 if and only if θ_0 lies outside the confidence interval. Use two-sided confidence intervals for two-tailed tests and one-sided confidence intervals for one-tailed tests. The level of the test is α. The power of the test would depend on the alternative value of θ being considered and the sample size n. The width of the confidence interval is related to the power of the test. The narrower the confidence interval is, the more powerful the test will be (at any alternative).

On the other hand, suppose you have an α-level test of the null hypothesis H_0: $\theta = \theta_0$ versus the alternative H_1: $\theta \neq \theta_0$, then a $100(1 - \alpha)\%$ bootstrap confidence interval is given by the set of estimates (or their corresponding test statistics) that fall between the upper and lower critical values for the test. Good (1994) illustrated this for the location parameter (mean = median) of a uniform distribution.

In the previous chapter, we saw many ways to compute bootstrap confidence intervals. For each confidence interval method that we can produce, we can construct a corresponding hypothesis test on the parameter. The null hypothesis is rejected if and only if the parameter value or values associated with the null hypothesis fall outside the confidence interval.

Let us consider an example where the parameter is the ratio of two variances. We use this ratio to test for equality of variances. So the null hypothesis is that the ratio equals 1. We provide this example to illustrate how bootstrap tests can be obtained from confidence intervals.

Fisher and Hall (1990) showed that for the one-way analysis of variance, the F-ratio is appropriate for testing equality of means (even for non-normal data) as long as the variance in each group is the same. This is because for equal variances, the F-ratio is asymptotically pivotal. Since bootstrap confidence intervals have nice properties when the statistic involved is asymptotically pivotal, the corresponding hypothesis test will work well also.

In the case when the variances differ significantly, the F-ratio is not asymptotically pivotal, and hence, Hall and Fisher used a statistic originally proposed by James (1951) that is asymptotically pivotal. The problem is that the ratio of the variances now depends on the variances themselves which become nuisance parameters, whereas before, all those ratios could be assumed to be equal to 1.

In an unpublished paper (Good and Chernick, 1993), an F-ratio was used to test equality of two variances. Under the null hypothesis, the two variances are equal and the ratio of the standard deviations is 1. So there is no nuisance parameter for the F-ratio and it is pivotal.

In section 3.3.2 of Good (1994), he pointed out that exact permutation tests cannot be devised for this problem when the sample sizes $n1$ and $n2$ are not equal (the exchangeability assumption breaks down). But it is easy to bootstrap such a sample by bootstrapping at random with replacement $n1$ times for population 1 and $n2$ times for population 2. Under the null hypothesis, the F-ratio should be close to 1. Using the Monte Carlo approximation to the bootstrap, we find a test simply by choosing a corresponding bootstrap confidence interval for the F-ratio and determining whether or not the interval contains 1. This is the right approach to take for confidence intervals but not for hypothesis testing.

EXAMPLE 4.1   Two-Sample Bootstrap

The following unoptimized R script segment illustrates a method of resampling two different samples:

```
> set.seed(1)
    > n1<- 29
    > x1<- rnorm(n1, 1.143, 0.164) #some random normal
variates
```

```
> n2<- 33

> x2<- rnorm(n2, 1.175, 0.169) #2nd random sample

> theta<- as.vector(NULL) #vector to hold difference
estimates

> for (i in 1:1000) { #bootstrap resamples

+ xx1<- sample(x1, n1, replace=TRUE)

+ xx2<- sample(x2, n2, replace=TRUE)

+ theta[i]<- mean(xx1)-mean(xx2)

+ }

> quantile(theta, probs=c(.025,0.975)) #Efron
percentile CI on difference in means

  2.5% 97.5%

-0.1248539 0.0137601
```

The 95% confidence interval contains the value 0.0, so we cannot reject H_0: $\mu_1 = \mu_2$ versus H_0: $\mu_1 \neq \mu_2$ at the $\alpha = 0.05$ significance level.

4.2   WHY TEST HYPOTHESES DIFFERENTLY?

Our example ignores a key issue called centering. Our confidence interval is centered at the null hypothesis value of $\sigma_1^2/\sigma_2^2 = 1$. But Hall (1992, section 3.12) pointed out that for hypothesis testing, it is better in terms of power to center at the original sample estimate. The idea is that in bootstrapping, we should use the bootstrap version of the null distribution, which would be centered at the sample estimate. In the same spirit, the null distribution for the bootstrap samples would be to bootstrap $n1$ times for population 1 and $n2$ times for population 2 from the pooling of the two original samples. This again differs from what we would do for a confidence interval. For further discussions about this point, see Hall (1992) or Hall and Wilson (1991). This article and the section in Hall's book, referenced above, pointed out that inverting confidence intervals is not the only and perhaps not the best way to handle hypothesis testing.

The crucial difference between hypothesis testing and confidence intervals is the existence of the null hypothesis. Often, the null hypothesis is simple. By simple we mean that it contains only a single point in the parameter space. Often the point is zero as in the case of testing equality of means where the parameter is the difference between the two population means. On the other hand, for a confidence interval in this case, zero plays no role. We are just looking for a set of plausible values for the true mean difference.

Also, the P-value and significance level of the test are both related to the null hypothesis. They are probabilities conditional on the null hypothesis being correct. So to reject the null hypothesis, we want to see how extreme our test statistic would be when the null hypothesis is true. Because hypothesis testing involves how extreme the test statistic is under the null hypothesis, the reference distribution in hypothesis testing should be the pooled distribution. So in the case of the mean difference for two populations, bootstrap the full sample of size $n + m$ from the pooled data with replacement where n is the sample size for population 1 and m is the sample size for population 2. This is covered very briefly on page 67 of Chernick (2007). Refer to Hall (1992, pp. 148–151) and Hall and Wilson (1991) for a clear rationale and an explanation as to why this is different than how you get bootstrap confidence intervals (regions). Explicit prescriptions for doing this are given in Efron and Tibshirani (1993, pp. 220–224) and Davison and Hinkley (1997, pp. 161–164). The same argument is made in Shao and Tu (1995, pp. 176–177).

There is nothing wrong with inverting the confidence intervals for the mean difference even though the confidence interval samples separately from the two distributions. However, for testing, such tests may not be very powerful. Some similar situations have been confronted in elementary statistical problems. For example, suppose we are testing to see if a person is really psychic. Then we would provide a series of questions and see if the individual's performance is better than chance. Then, the null hypothesis is that $P = 1/2$ and the alternative is $P > 1/2$. Now we could do an exact binomial test but suppose that n is large enough for the normal distribution to be a good approximation to the binomial. If we were looking at a confidence interval, we would want to use $(p_h - p)/sqrt(p_h\{1 - p_h\}/n)$ for the pivotal quantity. However, for the hypothesis test, it is often recommended to use $(p_h - p)/sqrt(p_0\{1 - p_0\}/n)$, where p_0 is the point null hypothesis value. In this case, $p_0 = 1/2$. This is known to lead to a more powerful test as the normalizing of the test statistic is based on the null hypothesis.

4.3   TENDRIL DX EXAMPLE

In 1995, Pacesetter, Inc., a St. Jude Medical company that produces pacemakers and leads for patients with bradycardia, submitted a protocol to the U.S. Food and Drug Administration (FDA) to conduct a pivotal phase III trial to determine the safety and effectiveness of a steroid-eluting lead. The study was to compare the Tendril DX model 1388T (a steroid-eluting lead) with a concurrent control, the marketed Tendril DX model 1188T.

The two leads would be identical active fixation leads except for the different tips on the leads with the 1388T having a tip designed to allow steroid liquid to slowly drip out from it (titanium nitride on the tip of the 1388T along with the steroid-eluting plug). Both leads can be assigned placement in the atrium or the ventricle, and comparisons were made for each chamber of the heart. They were implanted with the Pacesetter Trilogy DR+ pulse generator, the top-of-the line dual chamber pacemaker at that time.

Medtronic had previously run a similar trial with its steroid-eluting leads and had gotten FDA clearance to sell the leads. Based on the results of that trial, Pacesetter had a lot of useful information to design a successful trial of their own. The steroid drug is known to reduce inflammation that occurs at the site where the lead is implanted. The greater the inflammation, the higher the capture threshold will be during the acute phase (usually considered to be the first 6 months after the implant).

Chernick proposed a 0.5-V average decrease in capture threshold for the steroid lead compared with the nonsteroid lead at the follow-up visit 3 months after implant. The sample size was selected to have at least 80% power to detect an improvement of 0.5 V at 3 months. Because Medtronic already had a steroid-eluting lead that demonstrated safety and efficacy to the satisfaction of the FDA, a 1:1 randomization though probably optimal from the statistical point of view would not seem very ethical since it would put half the patients on an expected inferior lead. So Pacesetter chose a 3:1 randomization in order to be ethical and to provide incentive for subjects to enroll in the study.

This same criterion was applied to the comparisons in both the atrium and the ventricle. Chernick determined that with a 3:1 randomization, 132 patients would be needed, 99 getting the steroid-eluting lead and 33 getting the control lead. For 1:1 randomization, the same power can be achieved with 50 patients in each group. So the cost is an additional 32 patients. The protocol was approved by the FDA. St. Jude Medical submitted interim reports and a premarket approval application. The leads were approved for market release in 1997.

Since the capture threshold takes on distinct values, Gaussian models would only be appropriate for averages in very large samples. So in addition to the standard t-test, the Wilcoxon rank sum test was used for the main test and bootstrap confidence intervals were inverted to get a bootstrap test (only the simple percentile method was used).

The results for the atrial chamber were as follows for the mean difference between capture thresholds at 3 months (control–treatment). Here, a positive difference indicates that the nonsteroid lead had a high capture threshold than the steroid lead. Lower thresholds are better and drain the pulse generator's battery more slowly (Table 4.1).

The results for the ventricle were as follows for the mean difference between capture thresholds at 3 months (control–treatment) (Table 4.2).

TABLE 4.1. Summary Statistics for Bootstrap Estimates of Capture Threshold Differences in Atrium Mean Difference Nonsteroid versus Steroid

No. in control group	No. in treatment group	No. of bootstrap samples	Estimated mean difference	Estimated standard deviation	Minimum difference	Maximum difference	Percentage of differences below 0.5 V
29	89	5000	1.143	0.164	0.546	1.701	0%

TABLE 4.2. Summary Statistics for Bootstrap Estimates of Capture Threshold Differences in Ventricle Mean Difference Nonsteroid versus Steroid

No. in control group	No. in treatment group	No. of bootstrap samples	Estimated mean difference	Estimated standard deviation	Minimum difference	Maximum difference	Percentage of differences below 0.5 V
33	109	5000	1.175	0.169	0.667	1.951	0%

4.4 KLINGENBERG EXAMPLE: BINARY DOSE–RESPONSE

In drug discovery, an important part of phase II is the establishment of a dose–response relationship for the drug. At the same time, it is also important to establish a minimally effective dose and a maximally safe dose. This is important to establish a best dose or doses to test in phase III.

We now present results from Klingenberg (2007). The purpose of the paper is to show how new adaptive design methodology can speed up the drug development process. Also, failure in phase III often happens because the best dose is not chosen. Failure in phase III has been increasing and was approaching 50% in 2007. Such failures are extremely costly for the pharmaceutical companies.

Klingenberg proposed combining proof of concept and dose–response in an adaptive phase IIa–IIb trial. Adaptive designs hold great promise for speeding up drug development and have had some demonstrated success using either frequentist or Bayesian approaches. See Chow and Chang (2006) for more details on the various types of adaptive design trials and the reasons for their implementation. The interest and success of these adaptive approaches is even stronger in 2010, and the FDA is encouraging their use through a draft guidance document that was released in February 2010.

The idea of Klingenberg's approach is to use the following strategy: (1) along with clinical experts, identify a reasonable class of possible dose–response models; (2) pick from the list a subset that best describe the observed data; (3) use model averaging to estimate a target dose; (4) identify the models that show a significant signal (trend) establishing proof of concept; (5) use the permutation distribution of the maximum penalized deviance to pick the best model denoted s0; and (6) use the "best" model to estimate the minimum effective dose (MED). An important role in this procedure is the use of permutation methods to adjust P-values because of multiple testing.

Advantages of the approach over standard methods are that it (1) incorporates model uncertainty in the proof of concept decision and the estimation of a target dose, (2) yields confidence intervals for the target dose estimate, (3) allows for covariate adjustments, and (4) can be extended to more complicated data structures. Klingenberg also asserted that it is at least as powerful at detecting a dose–response signal as other approaches. He illustrated his method on a phase II clinical trial for a treatment for irritable bowel syndrome (IBS).

The bootstrap enters the picture as a procedure for determining confidence intervals for the "best" dose. The unified approach of Klingenberg is similar to the approach taken by Bretz et al. (2005) for normally distributed data. However, Klingenberg is applying it to binomial data.

IBS is a disease that reportedly affects up to 30% of the U.S. population at sometime during their lives (American Society of Colon and Rectal Surgeons http://www.fascrs.org). Klingenberg took data from a phase II clinical trial of a compound used to treat IBS. The target dose in the trial was based on a consensus of experts. Klingenberg reanalyzed these data using his method. Further details can be found in Klingenberg (2007) as well as Chernick (2007). Table 4.3 describes the models and Table 4.4 provides the results.

TABLE 4.3. Dose–Response Models Used to Determine Efficacy of Compound for IBS

Model ID	Type of model	Link function	Predictor	No. of permutations
M_1	Logistic	logit	$\beta_0 + \beta_1 d$	2
M_2	Log–log	log–log	$\beta_0 + \beta_1 d$	2
M_3	Logistic in log–dose	logit	$\beta_0 + \beta_1 \log(d + 1)$	2
M_4	Log linear	log	$\beta_0 + \beta_1 d$	2
M_5	Double exponential	identity	$\beta_0 + \beta_1 \exp(\exp(d/\max(d)))$	2
M_6	Quadratic	identity	$\beta_0 + \beta_1 d + \beta_2 d^2$	3
M_7	Fractional polynomial	logit	$\beta_0 + \beta_1 \log(d + 1) + \beta_2/(d + 1)$	3
M_8	Compartment	identity	$\beta_0 + \beta_1 d \exp(-d/\beta_2), \beta_2 > 0$	3
M_9	Square root	logit	$\beta_0 + \beta_1 d^{1/2}$	2
M_{10}	Emax	logit	$\beta_0 + \beta_1 d/(\beta_2 + d), \beta_2 > 0$	3

TABLE 4.4. G_s^2-Statistics, P-Values, Target Dose Estimates, and Model Weights

Model ID	Type of model	G_s^2	Raw P-value	Adjusted P-value	MED (mg)	Model weight (%)
M_1	Logistic	3.68	0.017	0.026	N/A	0
M_2	Log–log	3.85	0.015	0.024	N/A	0
M_3	Logistic in log–dose	10.53	$<10^{-3}$	0.001	7.9	6
M_4	Log linear	3.25	0.022	0.032	N/A	0
M_5	Double exponential	0.90	0.088	0.106	N/A	0
M_6	Quadratic	6.71	0.005	0.005	7.3	1
M_7	Fractional polynomial	15.63	$<10^{-4}$	$<10^{-4}$	0.7	81
M_8	Compartment	11.79	$<10^{-3}$	$<10^{-3}$	2.5	12

Critical value 2.40 MED(avg.) = 1.4 95%, confidence interval (CI) = (0.4, 12.0].
N/A, not applicable.

Table 4.4 above gives results for the models when only the first eight were considered. The model averaging gave 0 weight to models with low G_s^2-statistics (M_1, M_2, M_4, and M_5).

4.5 HISTORICAL NOTES

Discussion of bootstrap hypothesis tests appears in the paper of Efron (1979). Subsequent work can be found in Beran (1988), Hinkley (1988), Fisher and Hall (1990), and Hall and Wilson (1991). Some specific applications and simulation studies for bootstrap hypothesis testing were conducted by Dielman and Pfaffenberger (1988), Rayner (1990a, b), and Rayner and Dielman (1990).

It is Fisher and Hall (1990) who first pointed out that in spite of the close connection between hypothesis tests and confidence intervals, there are important differences that lead to special treatment for hypothesis tests. Particularly important in hypothesis testing is the use of asymptotically pivotal statistics and centering the distribution under the null hypothesis.

4.6 EXERCISES

1. Consider the alflatoxin residue data of Chapter 1 (Exercise 1.6.5): United States Environmental Protection Agency (USEPA) deems peanut butter adulterated if the mean aflatoxin residue is 20 ppb or more. The industry average for peanut was found to be 5.7 ppb in 1990 by consumer reports.
   (a) Estimate an Efron percentile bootstrap 90% confidence interval on the mean aflatoxin residue. Use $B = 1000$ resamples.
   (b) Compare the alfatoxin level found with the industry average value of 5.7 ppm: Is the upper confidence limit less than 5.7 ppb, or is it equal or above? What does this imply about a hypothesis test of H_0: $\mu > 5.7$ ppb versus H_1: $\mu < 5.7$ ppb at the $\alpha = 0.05$ significance level?
   (c) Find the P-value for the test in Exercise 4.6.1b. (Hint: Find the probability in the resampling distribution of getting 5.7 ppb or above.)

2. Consider the microbiological method data of Chapter 3 (Exercise 3.9.6):
   (a) Estimate an Efron percentile bootstrap 90% confidence interval on the mean difference in counts of the two methods. Use $B = 1000$ resamples.
   (b) Use the confidence interval in Exercise 4.6.2a to conduct a hypothesis test of H_0: $\mu_1 = \mu_2$ versus H_1: $\mu_1 \neq \mu_2$ at the $\alpha = 0.05$ significance level. Is there reason to believe the two methods are not equivalent?
   (c) Find the P-value for the test in Exercise 4.6.2b. (Hint: Find the probability in the resampling distribution of getting 0.0 as a result and adjust [double smaller probability] to get the two-tailed P-value.)

3. Information retrieval performance: When preparing for complex cases, lawyers use search services to find cases in the legal literature, which are relevant to the point in issue. The legal literature is a vast collection of case law and other materials, and the number of relevant documents in this database is typically small. "Recall" (R) denotes the proportion of all relevant documents that are actually retrieved (similar to "Sensitivity" in clinical studies). "Precision" (P) is the proportion of relevant documents retrieved to the total number retrieved (similar to "Positive Predictive Value" in clinical studies). High recall is important, so that key cases are not missed in the search. High precision is important, so that reviewing search results is not overly burdensome. A good search is one in which both R and P are high, but the two are typically inversely related. A commonly used combination figure of merit is F_2 defined by (see, e.g., http://en.wikipedia.org/wiki/Precision_and_recall)

$$F_2 = (5PR)/(4P + R) = (PR)/(0.8P + 0.2R).$$

A law firm wants to use a search service to research the issues in a large class action suit. Before they engage the (expensive) service, the law firm wishes to validate that $F_2 > 0.40$ at the 0.05 significance level. This is tested by running a pilot study of a small-scale search. It was decided to examine $N = 7478$ documents chosen at random from the full database. Expert reviewers went through each and every document, and found that $n = 150$ of these were actually relevant to the issue in point. The search program was run on the sample dataset, and a total of $m = 748$ were allegedly found to be relevant, with the actual number relevant actually $a = 123$. The resulting 2×2 matched classification table ("confusion matrix") was as follows:

		True status		Total
		Relevant	Irrelevant	
Search	Retrieved	$a = 123$	625	$n = 748$
Program	Not retrieved	27	6703	6730
Total		150	7328	$N = 7478$

(a) Find the observed Recall R, Precision P, and the figure of merit F_2.
(b) Resample the 2×2 contingency table $B = 1000$ times. (Hint: Use the multinomial distribution and rmultinom() in R.)
(c) Find 90% and 95% confidence intervals for the true F_2 for the complete database using Efron's percentile method.
(d) Conduct a test at the 0.05 significance level of $H_0: F_2 \leq 0.4$ versus $H_1: F_2 > 0.4$. Should the search service be engaged for this lawsuit?

(For more discussion of this type of text mining problem, see Dabney, 1986.)

REFERENCES

Beran, R. J. (1988). Prepivoting test statistics: A bootstrap view of asymptotic refinements. J. Am. Stat. Assoc. 83, 687–697.

Bretz, F., Pinheiro, J., and Branson, M. (2005). Combining multiple comparisons and modeling techniques in dose-response studies. Biometrics 61, 738–748.

Chernick, M. R. (2007). Bootstrap Methods: A Guide for Practitioners and Researchers, 2nd Ed. Wiley, Hoboken, NJ.

Chow, S. C., and Chang, M. (2006). Adaptive Design Methods in Clinical Trials. Chapman and Hall/CRC Press, Boca Raton, FL.

Dabney, D. P. (1986). The curse of Thamus: An analysis of full-text legal document retrieval. Law Library Journal 78, 5–40.

Davison, A. C. and Hinkley, D.V. (1997). Bootstrap Methods and Their Application. Cambridge University Press, Cambridge.

Dielman, T. E., and Pfaffenberger, R. C. (1988). Bootstrapping in least absolute value regression: An application to hypothesis testing. Commun. Stat. Simul. Comput. 17, 843–856.

Efron, B. (1979). Bootstrap methods: Another look at the jackknife. Ann. Stat. 7, 1–26.

Efron, B., and Tibshirani, R. (1993). An Introduction to the Bootstrap. Chapman and Hall, New York.

Fisher, N. I., and Hall, P. (1990). On bootstrap hypothesis testing. Aust. J. Stat. 32, 177–190.

Good, P. (1994). Permutation Tests. Springer-Verlag, New York.

Good, P., and Chernick, M. (1993). Testing the equality of variances of two populations. Unpublished manuscript.

Hall, P. (1992). The Bootstrap and Edgeworth Expansion. Springer-Verlag, New York.

Hall, P., and Wilson, S. R. (1991). Two guidelines for bootstrap hypothesis testing. Biometrics 47, 757–762.

Hinkley, D. V. (1988). Bootstrap methods (with discussion). J. R. Stat. Soc. B 50, 321–337.

James, G. S. (1951). The comparison of several groups of observations when the ratios of the population variances are unknown. Biometrika 38, 324–329.

Klingenberg, B. (2007). A unified framework for proof of concept and dose estimation with binary responses. Submitted to Biometrics.

Rayner, R. K. (1990a). Bootstrapping p-values and power in the first-order autoregression: A Monte Carlo investigation. J. Bus. Econ. Stat. 8, 251–263.

Rayner, R. K. (1990b). Bootstrap tests for generalized least squares models. Econom. Lett. 34, 261–265.

Rayner, R. K., and Dielman, T. E. (1990). Use of bootstrap in tests for serial correlatiomn when regressors include lagged dependent variables. Unpublished Technical Report.

Shao, J., and Tu, D. (1995). The Jackknife and Bootstrap. Springer-Verlag, New York.

5

TIME SERIES

5.1 FORECASTING METHODS

A common and important problem is forecasting. We try to forecast tomorrow's weather or when the next big earthquake will hit. But our efforts are fraught with uncertainty. When there exist historical data and we can find empirical or mechanistic models to fit to these data, we may be able to produce accurate forecasts. For certain problems such as the Dow Jones Industrial Average (DJIA) and for earthquakes, good models are not available to fit to the historical data, and forecasting becomes very problematic. In some cases such as the DJIA, there is a good empirical time series model called the random walk. But that model leaves too much unexplained variability to be useful in forecasting.

With the exception of finance that requires more volatility, commonly used forecasting techniques based on historical data without covariates are the Box–Jenkins autoregressive integrated moving average (*ARIMA*) models and exponential smoothing. The simplest form of exponential smoothing is a particular *ARIMA* model namely the *IMA*(1, 1) model. The basic approach for iteratively selecting and fitting an *ARIMA* model was first developed in Box and Jenkins (1970). The text has been revised twice

An Introduction to Bootstrap Methods with Applications to R, First Edition. Michael R. Chernick, Robert A. LaBudde.
© 2011 John Wiley & Sons, Inc. Published 2011 by John Wiley & Sons, Inc.

since 1970 (Box and Jenkins, 1976; Box et al., 1994). The autoregressive models that are an important subclass of the *ARIMA* models go back to Yule (1927).

The basic exponential smoothing model provides forecasts of future values using exponentially decreasing weights on the previous observations to produce a weighted average for the forecast. In single exponential smoothing, the weights are determined by a single parameter called the smoothing constant. As previously mentioned, this form of exponential smoothing is the *IMA*(1, 1) process.

The notation for the general *ARIMA* process is *ARIMA*(p, d, q) where p is the number of autoregressive parameters in the model, d is the number of times differencing is used to achieve stationarity, and q is the number of moving average parameters in the model. Differencing works as follows. We transform the time series X_t to Z_t where $Z_t = X_t - X_{t-1}$. Informally, stationarity is determined by the sample autocorrelation function.

For stationary time series, the autocorrelation function should rapidly decay to 0. If it stays close to 1 in absolute values for many lags (previous time periods), it is best treated as a nonstationary process, and differencing is done to see if the resulting process is stationary. If it is still nonstationary, another difference is tried, and this differencing procedure is continued until the transformed time series appears to be stationary. So for simple exponential smoothing, $p = 0$, $d = 1$, and $q = 1$. So there is only one moving average parameter that determines the smoothing constant.

5.2   TIME DOMAIN MODELS

ARIMA models are attractive because they empirically provide a good approximation to the observed time series for many stationary and nonstationary processes. The theory of stationary time series provides the answer. Stationary stochastic processes have both an infinite moving average representation and an infinite autoregressive representation. This means that a finite moving average or finite autoregression will provide a good approximation to almost any stationary process with finite second moments but possibly with a large number of parameters.

By allowing both autoregressive and moving average terms to appear in the model, a representation that is parsimonious in terms of parameters can be constructed. The first difference operator (transformation) allows for nonstationary processes containing a linear trend to be included. High-order differencing allows for quadratic, cubic, and higher-order polynomial trends. Sometimes a time series is nonstationary because it has seasonal cycles. In that case, seasonal trends are removed by taking seasonal differences. This will remove periodic functions from the time series. The *ARIMA* models get generalized to handle data with cycles. These are the *ARIMA* plus seasonal trend models.

Although the Box–Jenkins models cover a large class of time series and provide very useful forecasts and prediction intervals as well as a systematic method for model building, they do have some drawbacks. Parameter estimation is based on maximum likelihood under the assumption of Gaussian residuals. If residuals depart markedly from the Gaussian distribution, these estimates may not be very good.

This is often seen when the residual distribution is heavy tailed or the time series has some outliers (even if for only a small percentage of the data). Also, the methodology that relies on the sample autocorrelation and partial autocorrelation to make decisions about the order of the models (i.e., appropriate values for p, d, and q) falls apart in the presence of outliers because the estimates of these functions are very sensitive to outliers. This is discussed in Chernick et al. (1982) and Martin (1980).

A remedy for this problem is to detect and remove the outliers and then fit a Box–Jenkins model with the outlying observations missing. Another approach is to use robust parameter estimates. In some approaches, observations are all included but down-weighted when they are extreme. As Huber pointed out for location estimates, this can be similar to removing outliers except that it is less abrupt.

Outlier removal can be viewed as giving 0 to observations outside the detection threshold and equal weight to all observations within the detection threshold (in the location parameter case). Although this idea does not formally generalize to more complicated situations, it shows that certain types of robust estimates are like smoothed out generalizations of outlier removal. The work of Rouseeuw and Leroy (1987) is one of several good texts on robust methods.

In the 1980s, there were interesting theoretical developments in bilinear and nonlinear time series models that may be useful for time series analysts in dealing with data that do not fit well to the Box–Jenkins models. See Tong (1983, 1990) for this theory. Also, because of the high variability in the time series (when the variance changes significantly, over time it is called volatility by economists), many financial models were needed to deal with the problem.

The list of such financial models include the autoregressive conditional heteroscedastic (*ARCH*) models, their generalization (*GARCH* and exponential *GARCH*) conditional heteroscedastic autoregressive moving average (*ARMA*) (*CHARMA*) models, the random coefficient autoregressive models, and the stochastic volatility (*SV*) models. For a detailed description with their advantages and disadvantages, see Pena et al. (2001, chapter 9) or the original articles by Engel (1982), Bollerslev (1986), Tsay (1987), Nelson (1991), Nicholls and Quinn (1982), Melino and Turnbull (1990), and Harvey et al. (1994).

Other time series approaches in the time domain, include nonparametric time series models and neural network models. The point of mentioning these nonlinear, nonparametric, and heteroscedastic models is that although the bootstrap does not play a role when the *ARIMA* models work well, it does have a role when more complex time domain methods are needed. Another book by Weigend and Gershenfeld (1994) covers a lot of advanced methods for forecasting with time series and several include bootstrap methods. Also, as we shall see in the next section, when a simple *AR* model is appropriate but Gaussian assumptions are not, the bootstrap can help in producing more accurate prediction intervals.

5.3 CAN BOOTSTRAPPING IMPROVE PREDICTION INTERVALS?

For stationary autoregressive processes, there are some results regarding the improvements possible using bootstrap intervals as opposed to the standard ones involving

Gaussian assumptions. To illustrate this, we shall look at the simplest case, the $AR(1)$ model. The form of the model is as follows:

$$y_t = b_1 y_{t-1} + e_t,$$

where y_t is the observed value of the time series at time t, e_t is the innovation or random noise component at time t that is independent of the past, and b_1 is the first-order autoregressive parameter that relates y_t to y_{t-1}. The process could be centered to have zero mean. This centering can be done by taking the average of the y's over all observed time points and subtracting it from each observation.

For Gaussian processes, the maximum likelihood estimate for b_1 is computed along with the standard error of the estimate. If y_{tm} is the last observation, then the one-step-ahead forecast for y_{tm+1} is simply obtained as $\hat{b}_1 y_{tm}$ where $\hat{b}_1$ is the maximum likelihood estimate of b_1. Statistical software packages including SAS/ETS, SPSS, Autobox, and Minitab are all capable of doing these calculations to compute parameter estimates, their standard errors, and the prediction point and interval estimates.

These programs provide reasonable prediction intervals if the innovation process is approximately Gaussian with mean zero. Stine (1987) provided forecasts and prediction intervals with time series generated from the Gaussian model using the procedure just described. But he also provided a bootstrap approach and showed that although the bootstrap intervals are not as efficient as the ones produced by the standard $ARIMA$ software, they are much better when the innovations depart from the Gaussian distribution significantly.

The method for bootstrapping in the $AR(1)$ model is to bootstrap residuals just as we did in the model-based approach for regression. First, we obtain an estimate of b_1, which could be the maximum likelihood estimate. We then generate estimated residuals:

$$\hat{e}_t = y_t - \hat{b}_1 y_{t-1} \text{ for } t = 2, 3, \dots, tm.$$

Note that we cannot compute a residual $\hat{e}_1$ since there is no value for y_0. A bootstrap sample $y_2^*, y_3^*, \dots, y_{tm}^*$ is then generated by the following recursion after a bootstrap sample is taken from these residuals. To bootstrap the residuals, we just sample with replacement tm times from $\hat{e}_t$ for $t = 2, 3, \dots, tm$. This gives us $e_2^*, e_3^*, \dots, e_{tm}^*$, and we get

$$y_2^* = e_2^*, y_3^* = \hat{b}_1 y_2^* + e_3^*, \dots, y_{tm}^* = \hat{b}_1 y_{tm-1}^* + e_{tm}^*.$$

In a slightly different approach, Efron and Tibshirani (1986) took $y_1^* = y_1$ and $y_2^* = \hat{b}_1 + y_1^* e_2^*$. In the Monte Carlo approximation to the bootstrap, the e_i^*'s are repeatedly generated B times and the y_i^*'s are obtained as given above in each of the B bootstrap samples. In general for the pth order autoregressive process, we will need p initial values. Stine (1987) and Thombs and Schucany (1990) provided reasonable but different ways to get those initial p points in the series.

Bootstrap estimates of b_1 are obtained each of the B times by using the procedure that was used to get the original estimate of b_1 but using the sample $y_1^*, y_2^*, \dots, y_{tm}^*$. This is easily generalized to the $AR(p)$ model and also to general $ARMA(p, q)$ models. Efron and Tibshirani (1986) illustrated this using the Wolfer sunspot data and fitting an $AR(2)$ model.

Prediction intervals in the $AR(1)$ case depend on the unknown parameter b_1 and naturally would be less accurate when b_1 is replaced by an estimate. This would also be true for the mean square error (MSE) of the prediction estimate. Stine (1987) produced an estimate of this MSE when b_1 is estimated using the Taylor series. This estimate is good when the innovation component of the model is Gaussian. But in non-Gaussian cases, Stine (1987) provided a bootstrap estimate of the MSE that although biased, it is used to obtain more accurate predictions.

We now describe the approach of Stine, which we would advocate for these models when the innovations are non-Gaussian. Stine made the following assumptions:

1. Innovations have a mean of 0.
2. Innovations have finite second moments.
3. Innovations are symmetric about 0.
4. Cumulative distribution is strictly increasing.

While these are slight restrictions, 1, 3, and 4 are very reasonable. Condition 2 excludes innovation distributions with heavy tails. These conditions are satisfied by Gaussian innovations and generalize them in a sensible way. Efron and Tibshirani (1986) differ from Stine (1987) in that they do not assume property 3. Stine created the following innovation distribution with the above properties:

$$F_T = 1/2 + [L(x)/\{2(T-p)\}] \text{ for } x \geq 0$$
$$= 1 - FT(-x) \text{ for } x < 0,$$

where $L(x)$ = number of t for which $k\hat{e}_t \leq x$, for $t = p + 1, p + 2, \ldots, T$, and $k = [(T-p)/(T-2p)]^{1/2}$. This type F_T has the property that it produces bootstrap residuals that are symmetric about 0 and have the same variance as the original residuals. For more details on Stine's approach, see Stine (1987) or Chernick (2007, pp. 100–103).

Thombs and Schucany (1990) used a time reversal property to fit the last p observations and generate bootstrap samples in the case of a pth order autoregressive process. It should be pointed out that time reversibility, which applies to autoregressive Gaussian process, does not apply generally.

Chernick et al. (1988) showed that the exponential $AR(1)$ process of Gaver and Lewis is not time reversible, although its reversed version also has exponential marginal and is also a Markov process. There is not one preferred method for bootstrapping AR processes. These approaches are called model based and results depend heavily on the correct specification.

When a parametric model cannot be relied on, other approaches are needed. The most common is the block bootstrap, which was introduced in Kunsch (1989). The idea is to resample blocks (intervals of a time series) with replacement. The blocks are usual disjoint and cover the entire time series, but they can be overlapping.

5.4 MODEL-BASED METHODS

We have already introduced several time series models that are based on equations that express the current observation as a function of previous observations along with a random noise component. We have discussed in some simple cases (e.g., the $AR(1)$ model) how we can bootstrap residuals to get bootstrap estimates of parameters and repeat to get a bootstrap distribution for the parameter estimates. These methods all fall into the category of model-based methods.

If the model exactly described the process and we actually knew the exact values of the parameters, we could invert the equation to express the noise terms exactly. This gives us the independent sequence of random variables that we knew how to bootstrap in Chapter 1. Since in practice we do not know the parameters, we estimate them and then estimate the noise using the equation that expresses each noise term as a function of the past observations. This sequence has a weak dependence because common estimates of parameters are involved. But just as with regression problems, this near independence makes bootstrapping residuals a useful approach.

In the case of the regression model, we saw that misspecification in the model can create biases and cause the model-based method to fail. In that case, bootstrapping vectors provided a solution that was not sensitive to model misspecification. The same is true for time series models. The block bootstrap methods were touched on briefly in the previous section. The various block bootstrap methods will be described in more detail in the next section.

5.4.1 Bootstrapping Stationary Autoregressive Processes

One of the most highly studied model-based methods is the bootstrapping of stationary autoregressive processes. We shall adopt the term autoregressive bootstrap and the notation autoregressive bootstrap (ARB) as described in Lahiri (2003, section 8.2).

The pth order stationary autoregressive process is defined by

$$X_i = \beta_1 X_{i=1} + \beta_2 X_{i-2} + \ldots + \beta_p X_{i-p} + e_i, \tag{5.1}$$

where the p β_j's ($j = 1, 2, \ldots, p$) are the autoregressive parameters and the e_i's form a zero-mean independent and identically distributed (i.i.d.) sequence of random variables with common distribution function F.

It is known (see Box and Jenkins, 1976 or Brockwell and Davis, 1991) that the β_j's must satisfy certain conditions on a polynomial function called the characteristic polynomial for stationarity to hold. The characteristic polynomial is a polynomial with real coefficients and hence for any degree k, it has k roots (some may be repeated) that lie in the complex plane. The stationarity condition requires all k roots to fall strictly inside the unit circle in the complex plane.

When at least one root lies on or outside the unit circle, the time series is nonstationary, and when all roots lie strictly outside the unit circle, it is sometimes called explosive. In the nonstationary case, there are ways to bootstrap in order to estimate the parameters, and we will deal with those cases later. The characteristic polynomial for the $AR(p)$ process given in Equation 5.1 is as follows:

$$\Psi_p(z) = z^p - \beta_1 z^{p-1} - \beta_2 z^{p-2} - \ldots - \beta_{p-1} z - \beta_p. \tag{5.2}$$

In general, this polynomial has p roots (not necessarily distinct) that all lie in the complex plane.

Another well-known property of stationary autoregressive processes is that they have an infinite moving average representation. This representation is sometimes useful in developing theories. The bootstrap approach for the $AR(p)$ model is just as previously described. However, there is an alternative approach in the literature due to Datta and Sriram (1997).

Bose (1988) investigated the properties of ARB when the normalized least squares estimates of the autoregressive parameters are used. Under the conditions described in Theorem 1, Bose showed that ARB is strongly consistent. This theorem is stated as theorem 8.1 in Lahiri (2003):

Theorem 2   Assume, for an $AR(p)$ process as defined in Equation 5.1, that the sequence $\{e_i\}\ i = 1, 2, 3, \ldots, \infty$ of i.i.d. random variables such that (1) $E(e_1) = 0$, $E(e_1^2) = 1$, and $E(e_1^8) < \infty$, and (2) the vector $(e_1, e_1^2)^T$ satisfies Cramer's condition, which states $\limsup \left| E\left(\exp\left(i\left(e_1, e_1^2\right)t\right)\right)\right| < 1$, where $i = \sqrt{-1}$, $\|t\| \to \infty$. Then if we also assume that the characteristic polynomial given by Equation 5.2 has all its roots inside the unit circle, $\sup_{x \in R_p} \left| P^*\left(n^{1/2}\Sigma^{1/2}\left(\beta_n^* - \beta_n\right) \le x\right) - P\left(n^{1/2}\Sigma^{1/2}\left(\beta_n - \beta\right) \le x\right)\right| = o\left(n^{-1/2}\right)$, almost surely.

The proof of this theorem is given in Bose (1988). As this is a technical probability result, let us describe what it says.

This theorem tells us that if our $AR(p)$ process is stationary and the innovation terms have a finite eighth moment and satisfy a technical condition called Cramer's condition, the ARB estimate of the p-dimensional vector β_n is strongly consistent for β. That means that for any x in p-dimensional Euclidean space, the probability distribution for the normalized difference between β_n^* and $\hat{\beta}_n$ under bootstrap sampling approaches the probability distribution under the $AR(p)$ model for the same normalized difference between $\hat{\beta}_n$ and β. So the bootstrap principle holds in the strict probability sense of almost sure convergence.

Because $\hat{\beta}_n$ is consistent for β, so too is the bootstrap estimate. It also indicates that since the rate of convergence is $o(n^{-1/2})$, ARB is second-order accurate under these conditions. The number n is used to denote the series has length n with n growing toward infinity. Σ is the $p \times p$ covariance matrix of β, and $\hat{\Sigma}$ is the bootstrap estimate of it. Since Σ is a non-negative definite matrix, there exists a symmetric matrix $\Sigma^{1/2}$ satisfying $\Sigma = \Sigma^{1/2}\Sigma^{1/2}$. $\Sigma^{1/2}$ is sometimes called a square root matrix for Σ. It need not be unique but any such matrix will do.

This theorem asserts second-order accuracy only when the X_i's have zero mean. This is a consequence of condition (1) in the theorem. However, in general, a stationary $AR(p)$ process can have a nonzero constant for its mean, and so in practice, we center the process by subtracting the sample mean of the X_i's. Bose pointed out that the result holds even in the more general case.

Lahiri (2003, p. 206) simulated two-sided equal tailed 90% and 80% confidence intervals showing ARB and moving block bootstrap (MBB) with four different lengths: $l = 2, 5, 10, 20$. This is a situation where ARB is theoretically better, which can be seen from how closely it comes to the confidence intervals for the true distribution of the normalized least squares estimator.

MBB is not very good for $l < 10$, but for l between 10 and 20, it gets reasonably close. So even though MBB suffers in this case, it is more robust to model misspecification and may be a better option in practice when you are uncertain about the correctness of the model. Choice of the length or block size in MBB is critical as we see here. We shall not cover this issue but the interested reader can see the details in Lahiri (2003, chapter 7).

EXAMPLE 5.1   Mechanically Separated Meat (MSM) Chicken Lots

After the removal of retail cuts (breast, thigh, leg, wing) from a chicken in fabrication post-slaughter, the body cavity portion of the carcass remains with small amounts of muscle tissue attached. These carcasses are accumulated and passed through machinery that efficiently separates residual muscle tissue from bone. The resulting by-product is denoted "MSM." In the production of MSM chicken, each lot collected is chemically analyzed for fat and classified by fat content. It is important that there be little or no autocorrelation of fat among lots produced, as this indicates drift in efficient operations. For this reason, it is interesting to determine the coefficient in an $AR(1)$ model along with its 95% confidence interval and a test for this coefficent being zero.

Using R and data for the fat content of 258 consecutive lots of nominal 18% fat MSM chicken paste,

```
> require('boot')
> c18<- read.table('c18.txt', header=TRUE)
> head(c18)
Fat
1  17.0
2  16.0
3  16.5
4  16.5
5  15.5
6  18.4
> tsc18<- ts(c18)  #convert to time series
object
  > a1<- acf(tsc18)  #autocorrelation function
```

```
> a1
Autocorrelations of series 'tsc18', by lag
0 1 2 3 4 5 6 7 8 9 10 11 12 13
1.000 0.092 0.150 0.166 0.095 0.098 0.185
0.138 0.084 0.079 0.151 0.069 0.002 0.107
14 15 16 17 18 19 20 21 22 23 24
0.040 0.019 0.068 0.032 0.104 0.045 0.006
-0.032 -0.092 0.027 0.051

> plot(a1)
> p1<- pacf(tsc18) #partial autocorrelation
function
> p1
Partial autocorrelations of series 'tsc18',
by lag
1 2 3 4 5 6 7 8 9 10 11 12 13 14
0.092 0.142 0.145 0.056 0.049 0.142 0.090
0.013 -0.002 0.095 0.013 -0.082 0.036 0.001
15 16 17 18 19 20 21 22 23 24
-0.023 0.006 -0.009 0.097 0.009 -0.053
-0.074 -0.108 0.022 0.050

> plot(p1)
> fit1<- arima(tsc18, order=c(1,0,0)) #fit
with AR(1)
> fit1
Call:
arima(x = tsc18, order = c(1, 0, 0))
Coefficients:
ar1 intercept
0.0918 17.3036
s.e. 0.0619 0.0954
sigma^2 estimated as 1.937: log likelihood =
-451.35, aic = 908.71
> confint(fit1) #95% CI on coefs
2.5 % 97.5 %
ar1 -0.02945246 0.2131153
intercept 17.11668786 17.4905042
```

The 95% parametric confidence interval on the $AR(1)$ coefficient is (-0.029, 0.213).

Now, find a bootstrap confidence interval using model-based (residual) resampling:

```
> ar1 <- ar(tsc18, order.max=1) #fit AR(1)
> armodel <- list(order=c(1,0,0),ar=ar1$ar)
> arres <- ar1$resid
> arres <- arres-mean(arres) #center
>
> bootf<- function (tsb) { #statistic for
time-series bootstrap
+ fit<- arima(tsb, order=c(1,0,0)) #fit with
AR(1)
+ return(coef(fit)[1])
+ }
> bootsim <- function(res, n.sim, ran.args)
{
+ # random generation of replicate series
using arima.sim
+ rg1 <- function(n, res) sample(res, n,
replace=TRUE)
+ ts.orig <- ran.args$ts
+ ts.mod <- ran.args$model
+ mean(ts.orig)+ts(arima.sim(model=ts.mod,
n=n.sim, rand.gen=rg1, res=as.vector(res)))
+ }
> boot1<- tsboot(tsc18, bootf, R=1000,
sim="model", n.sim=nrow(tsc18), orig.t=FALSE,
ran.gen=bootsim,
+ ran.args=list(ts=tsc18, model=armodel))
> mean(boot1$t[,1])
[1] 0.08712331
> quantile(boot1$t[,1],
probs=c(0.025,0.975)) #Efron percentile CI on
coefficient
2.5% 97.5%
-0.03271095 0.20972895
```

The bootstrap 95% confidence interval is (−0.033, 0.210), not much different from the parametric one above. In both cases, 0.0 is inside the interval, but there is a suggestion of a positive autocorrelation present.

5.4.2 Bootstrapping Explosive Autoregressive Processes

Explosive autoregressive processes of order p are simply nonstationary processes that have all the roots of the characteristic polynomial lying outside the unit circle. For explosive processes, the mean is not necessarily zero and, in some cases, may not exist. While a stationary $AR(p)$ process does not depend on the p initial values after a large number of observations, the behavior of an explosive process can crucially depend on the initial p observed values.

Surprisingly, Datta (1995) showed that under rather mild conditions, the ARB method is consistent for the conditional distribution of the normalized and studentized least squares estimates of β. The condition on the moments is that $E\{\log(1 + |e_{p+1}|)\} < \infty$. This allows the e_i's to have heavy-tailed distributions.

5.4.3 Bootstrapping Unstable Autoregressive Processes

An unstable $AR(p)$ process is one that has at least one root on the unit circle and others inside. The standard ARB method will not work well in these cases. Lahiri (2003) described a way to bootstrap unstable $AR(p)$ processes. Again, it is Datta (1996) who showed that bootstrap estimates based on a resample size n equal the length of the time series. But Datta was able to show consistency if the resample size m tends to infinity but at a slower rate than n. This is another example where the m-out-of-n bootstrap provides the needed remedy.

5.4.4 Bootstrapping Stationary *ARMA* Processes

It is not difficult to get a consistent model-based bootstrap estimate of the parameters in an *ARMA* model *ARMA*(p, q). It can be done by taking advantage of the facts that the finite parameter *ARMA*(p, q) process has both an infinite moving average and autoregressive representation. The details can be found in Lahiri (2003). The resulting bootstrap procedure is referred to as *ARMAB* by Lahiri. He presented it in detail. The method comes from the original paper of Kreiss and Franke (1992), who establish its consistency. But the second-order properties of *ARMAB* remain unexplored.

5.5 BLOCK BOOTSTRAPPING FOR STATIONARY TIME SERIES

Because of the drawbacks of model-based methods when the model is misspecified as shown in Bose (1988), alternative time domain approaches were sought, and the most successful has been the MBB. The idea of the block bootstrap was introduced by Carlstein (1986) and further developed by Kunsch (1989).

The key idea in the development of block bootstrapping is that for stationary time series, when observations are separated far enough, in time, they are nearly uncorrelated, so blocks separated far enough can be treated as exchangeable, if the time series has length n and can be factored as $n = bl$, where b and l are integers.

We can then break the time series into b disjoint blocks each of length l. We want l to be long enough to preserve the correlation structure of the time series. We could bootstrap by sampling with replacement the b blocks of length l each bootstrap realization pastes together to get total length n. There are many variants to this approach but the idea is basically the same.

One of Kunsch's ideas was to make the blocks overlap, which would allow for a larger number of blocks to choose from than with nonoverlapping blocks. Many advances have taken place since 2000 and are documented in the text by Lahiri (2003, chapter 2). The various types of block bootstrap approaches covered by Lahiri include (1) MBB, (2) nonoverlapping block bootstrap (NBB), (3) circular block bootstrap (CBB), (4) stationary block bootstrap (SBB), and (5) tapered block bootstrap (TBB) (very briefly).

We shall now explain how overlapping blocks can be constructed. Suppose we choose a fixed block length of four and for the first block we choose (y_1, y_2, y_3, y_4), the first four observations of the time series. The second block could be (y_2, y_3, y_4, y_5) and the third might be (y_3, y_4, y_5, y_6). We continue this pattern until we reach the end of the series. This gives us a large number of blocks, namely $n - 3$ where n is the length of the series. An unwanted result of this is that the first three observations ($l - 1$ in general where l is the block size) appear in fewer blocks than all the other observations.

Note that y_1 appears in only one block, y_2 in only two blocks, y_3 in only three blocks, and the rest of the y_i in four blocks. To overcome this effect on the initial $l - 1$ observations can be overcome by wrapping the data in a circle where the first observation appears right after the nth observation. This is the method of Politis and Romano (1992) called the CBB.

Block bootstrap methods have two drawbacks because piecing together resampled blocks does not perfectly mimic the correlation structure in the original series and can weaken or strengthen the dependency in neighboring blocks depending on whether or not adjacent blocks are close or far apart. Lahiri (2003) showed that the MBB fails to be consistent if the length of the time series obtained by pasting the blocks together is equal to the length n of the time series. However, as we will see in Chapter 8, the m-out-of-n bootstrap is a remedy that works here as well as in other situations.

There are other remedies to improve the properties of block bootstrap method. These include methods called postblackening and resampling groups of blocks (e.g., each resample consists of three or four blocks at a time). This means that the blocks are grouped first and then the groups are sampled. Both these methods are covered in detail in Davison and Hinkley (1997, pp. 397–398).

Another preferred method for overcoming the difficulty of MBB and NBB is the SBB. The SBB differs from other blocking schemes in that the length of each block is random rather than fixed. The block length L is sampled from a geometric distribution so it has the following form:

$$P\{L = j\} = (1 - p)^j p \quad \text{for} \quad j = 1, 2, 3, \ldots, \infty.$$

For this distribution, the average block length $l = 1/p$. This method was first proposed by Politis and Romano (1994). The problem with the geometric distribution is that it includes $j = 0$ to be a proper distribution, and although the time series has finite length n, the block length is unbounded. It appears that a simple remedy would be to truncate on the left at 1 and at the right at n. It then becomes necessary to normalize the probabilities to sum to 1.

Block resampling appears to have advantages over model-based resampling because it is not dependent on the specification of a parametric model. It therefore applies to a broad class of stationary properties. Other variations of block bootstrap and some theory can be found in Davison and Hinkley (1997, pp. 401–403). They also discussed the choice of block length and the underlying theory. Lahiri (2003) gave a very detailed and more recent account. An even more recent account of this can be found in Lahiri's article in the volume "Frontiers in Statistics" (Fan and Koul, 2006).

Davison and Hinkley (1997) illustrated block resampling using the Rio Negro river height time series. We present that example here. The main concern in the study was that there was a trend for heights of the river near the town of Manaus to rise over time due to deforestation. A test for an increasing trend in the time series was applied. Although there was some evidence of a trend, the statistical test was inconclusive.

The trend test that they used is based on a linear combination of the time series observations namely, $T = \Sigma a_i Y_i$, where for $i = 1, 2, 3, \ldots, n$, Y_i is the ith term in the sequence of the Rio Negro river's level at Manaus, and the coefficient a_i is given as

$$a_i = (i-1)\left(\left[1-\{(i-1)/(n+1)\}\right]^{1/2} - i\left[1-i/(n+1)\right]^{1/2}\right) \quad \text{for} \quad i = 1, 2, 3, \ldots, n.$$

The test based on this statistic is optimal for detecting a monotonic trend when the observations are i.i.d. But the Rio Negro data show a clear correlation between nearby observations, and the trend is not montonic. A smoothed version of the Rio Negro data clearly shows this (see Davison and Hinkley, 1997, fig. 8.9, p. 403).

Davison and Hinkley decided to use the same test statistic T described above but used the stationary bootstrap to get a more correct estimate of the variance of this statistic under the null hypothesis of no trend. In the example, $T = 7.908$, but is this statistically significantly different from zero when a time series has a dependence structure but not trend?

The asymptotic distribution of T is Gaussian with mean 0 but variance estimated by the stationary bootstrap. Davison and Hinkley (1997) actually compared the stationary bootstrap with a fixed block bootstrap where the average block size from the stationary bootstrap is compared with the fixed block bootstrap with block size equal to the nearest integer to the stationary bootstrap's average. Although the result depends on the block size the lowest reasonable estimate of the variance is about 45 when the entire series is used. This leads to a P-value of 0.12, which is not indicative of a strong trend.

EXAMPLE 5.2   MSM Chicken (Continued)
Using R for the block bootstrap:

```
> boot2<- tsboot(tsc18, bootf, R=1000, l=16, sim="fixed")
#block bootstrap with block length = 16
> mean(boot2$t[,1]) #mean estimate
[1] 0.07234437
> quantile(boot2$t[,1], probs=c(0.025,0.975)) #Efron
percentile CI on coefficient
2.5% 97.5%
-0.1729400 0.3282077
```

The block 95% confidence interval is (−0.173, 0.328), considerably wider than the model-based and parametric intervals, and the mean estimate 0.072 is smaller. Both "b" and "l" are relatively small here (16), so block resampling may not be reliable.

5.6   DEPENDENT WILD BOOTSTRAP (DWB)

The DWB was presented by Shao (2010). The DWB extends the notion of the wild bootstrap described in Chapter 2. The primary intent of the wild bootstrap is to find appropriate model parameter estimates and their standard errors in the case of a general heteroscedastic linear model. The main purpose of the DWB is to have a general procedure that can be applied to time series data with advantages over the the MBB and its various modifications. Since the theory developed for the MBB applies to equally spaced observations, it is nice that the DWB can easily be applied to equally spaced or irregularly spaced data. In practice, possibly because of missing observations or because the measurements are taken at randomly spaced time points, irregularly spaced observations are common.

The DWB method applies to stationary weakly dependent time series. Although Shao's paper focuses on time series and contrasts DWB to MBB in a time series setting, the method that can be adapted to handle spatial data is possible and the author plans to publish results for spatial data in a future article. Formally, the DWB is defined as follows.

Suppose the infinite stationary time series $\{X_i\}$ for $i \in \mathbb{Z}$ (the integers) has mean $\mu = E(X_i)$, the same for each i because of stationarity and covariance function $\gamma(k) = cov(X_0, X_k)$, then the DWB pseudo-observations are defined as

$$X_i' = \bar{X}_n + \left(X_i - \bar{X}_n\right)W_i \quad \text{for} \quad i = 1, 2, \dots, n,$$

where $\bar{X}_n = n^{-1}\sum_{i=1}^{n} X_i$ is the sample mean and $\{W_i\}_{i=1}^{n}$ are n auxiliary random variables with the following properties:

1. The W_i's are independent of the X_i's.
2. $E(W_i) = 0$ and $var(W_i) = 1$ for all i.

3. $\{W_i\}$ is a stationary time series with covariance function $cov(W_i, W_j) = a$ $([i - j]/l)$ and the function $a(\cdot)$ is a kernel function with bandwidth parameter l that depends on n. Shao also defined $K_a(x) = \int_{u=-\infty}^{\infty} a(u) e^{-iux} du$ for a real number x and $i = \sqrt{-1}$.

With this setup, let $T_n = \sqrt{n}(X_n - \mu)$. Shao asserted under a few minor added conditions that T_n is asymptotically $N(0, \sigma_\infty^2)$, where $\sigma_\infty^2 = \Sigma \gamma(k)$. Then, if we take

$$\bar{X}_n' = n^{-1} \sum_{i=1}^{n} X_i',$$

which is the DWB analogue of $\bar{X}_n$ and T_n^* the DWB analogue of T_n, then Shao gets $T_n^* = \sqrt{n}(\bar{X}_n' - \bar{X}_n)$, also asymptotically normal (follows the bootstrap principle) with the same limit as T_n. Shao showed this by proving asymptotic equivalence of the DWB variance estimator of $var^*(T_n^*)$, which is the variance under DWB sampling for T_n^* and the corresponding one for the TBB of Paparoditis and Politis (2001). Most of Shao's paper consists of technical probability results and numerical examples that we will not go into here. We simply note that in the numerical examples, two forms of DWB are used to compare with various choices for MBB and TBB. The two methods DWB1 and DWB2 differ in the choice of the kernel $a(\cdot)$. DWB1 uses the Bartlett kernel, and DWB2 uses a tapered kernel that Shao described.

5.7 FREQUENCY-BASED APPROACHES FOR STATIONARY TIME SERIES

Stationary Gaussian processes are characterized by their mean and covariance function. When the process is a time series, the covariance function parameterized by time lag is called the autocovariance function. Without any loss of generality, we can assume the mean is zero and then the autocovariance function uniquely determines the stationary Gaussian time series.

Normalized by dividing by the process variance, the autocovariance function is transformed to the autocorrelation function. The Fourier transform of the autocorrelation function is well defined for stationary processes and is called the spectral density function. While the autocorrelation function is a time domain function, the spectral density function is a function of frequencies. There is a 1-1 correspondence between the spectral density and the autocorrelation function for stationary processes with zero mean and variance one, and for Gaussian processes, either function characterizes the process.

When model estimation or forecasting is based on the autocorrelation function, the approach is called a time domain approach. When it is based on estimates of the spectral density function, it is called a frequency domain approach. Bloomfield (1976), Brillinger (1981), and Priestley (1981) gave detailed accounts of the theory and practice of the frequency domain approach. It is not our purpose to give a detailed account of the methods. Rather, our goal here is to look at how the bootstrap plays a role in the frequency domain approach to time series analysis.

An important theoretical result for the sample periodogram (a sample function estimate of the spectral density that is the Fourier transform of the sample autocorrelation function) is the following: When the spectral density function for a stationary time series is well defined, the sample periodogram evaluated at the Fourier frequencies ($\omega_k = 2\pi k/n$ where $k = 1, 2, 3, \ldots, n$ with n equal to the length of the time series) has each its real and imaginary parts approximately independent and normally distributed with mean zero and variance $ng(\omega k)/2$, where $g(\omega k)$ is the true spectral density at k. See Brillinger (1981) for the details.

We exploit the approximate independence to bootstrap the periodogram instead of the time series. The periodogram is easy to compute thanks to the fast Fourier transform (FFT). This lets us use data that are nearly independent and approximately Gaussian, which is easier to deal with compared with the dependent time domain data that may be non-Gaussian.

5.8   SIEVE BOOTSTRAP

Another time domain approach to apply the bootstrap to a stationary time series is the sieve bootstrap. We let P be the unknown joint probability distribution for an infinitely long time series where the infinite set of observations is denoted as $[X_1, X_2, X_3, \ldots, X_n, \ldots]$. In the special case where the observations are i.i.d., we take the empirical distribution function F_n for the first n observations, which are the ones that we get to see. Since F_n is the one-dimensional estimate of the distribution F common to each X_i based on the first n, the joint distribution of the first n observations is by independence the product F^n obtained by multiplying F by itself n times. A natural estimate for the joint distribution would be $(F_n)^n$.

However, in the case of dependence, the joint distribution is not the product of the one-dimensional marginal distributions. The idea of the sieve is to approximate this joint distribution by distributions that get closer to P as n increases. This is an abstract notion that we can make concrete by considering specific time series models.

We consider the large class of stationary time series models that admit an infinite autoregressive representation. Buhlmann (1997) created a sieve by approximating P by the joint distribution of a finite $AR(p)$ model. For any such stationary model, the approximation get closer to P as $p \to \infty$. Because of stationarity, each one-dimensional marginal distribution is the same and can be estimated from the series. Because of the finite autoregressive structure, P can be derived as a function of the marginal distribution F and the autoregressive parameters. For the details, see Buhlmann (1997).

Another choice for the sieve was introduced in Buhlmann (2002) and is based on varying the length of a Markov chain for a time series that is categorical. For a given time series model, there can be a number of possible sieves. Lahiri (2002), in discussing Buhlmann's paper, pointed out that in choosing a sieve, we are trading off the accuracy of the approximating distribution with its range of validity.

In the method of Buhlmann (1997), we have a sequence of autoregressive models with order p_n tending to infinity at a rate slower than n; namely, we require $n^{-1} p_n \to 0$ as both n and p_n tend to infinity. Given this finite autoregressive time series model, we fit the autoregressive parameters and approximate the joint distribution of the first n observations using the estimated parameters and the estimated marginal distribution. The details can be found in Lahiri (2003, pp. 41–43). This sieve is called the autoregressive sieve, and in the original paper, Buhlmann (1997) established the consistency for it.

5.9 HISTORICAL NOTES

The use of *ARIMA* and seasonal *ARIMA* models for forecasting and control was first popularized by Box and Jenkins in the late 1960s and formally in their classic text in Box and Jenkins (1970). The most recent update of that book is in Box et al. (1994).

A classic text covering both the time and frequency approaches to time series is Anderson (1971). A popular text used for graduate students in statistics is Brockwell and Davis (1991), which also covers both the time and frequency approaches to time series analysis. Another fine text in the category is Fuller (1976).

Tong's book and original article (Tong, 1990 and Tong, 1983, respectively) cover many aspects of nonlinear time series. Bloomfield (1976), Brillinger (1981), and Priestley (1981) are excellent texts that all focus on the frequency or spectral approach to time series. Hamilton (1994) is another major text on time series.

Braun and Kulperger (1997) did work on using the Fourier transform approach to bootstrapping time series. Bootstrapping residuals was described early in Efron (1982) in the context of regression. Efron and Tibshirani (1986) showed how bootstrapping residuals provides improved estimates of standard error for autoregressive parameters in their presentation of the *AR*(2) model for the Wolfer sunspot data.

Stine (1987) and Thombs and Schucany (1990) provided refinements to improve on the approach of Efron and Tibshirani. Empirical studies that provided applications for bootstrap prediction intervals are Chatterjee (1986) and Holbert and Son (1986). McCullough (1994) provided an application for bootstrap prediction intervals. Bose (1988) provided theoretical justification to the bootstrap approach to general stationary autoregressive processes using Edgeworth expansions.

Results for nonstationary (aka explosive) autoregressions appear in Basawa et al. (1989, 1991a, b). Difficulties with the bootstrap approach in estimating MSE in the context of stationary *ARMA* models were noted in Findley (1986).

The block bootstrap idea was introduced by Carlstein (1986), but the key breakthrough came in Kunsch (1989) and then followed up by many others. Developments in block bootstrapping are well chronicled in Davison and Hinkley (1997) and Lahiri (2003). The TBB was introduced by Paparoditis and Politis (2001, 2002) who showed that it gives more accurate estimates of variance-type level-2 parameters than MBB and CBB. This is mentioned very briefly by Lahiri (2003); so for details, the reader is referred to the original papers.

Shao and Yu (1993) showed that bootstrapping for the process mean can be applied in general for stationary mixing stochastic processes. Hall and Jing (1996) applied resampling in more general dependent situations, and Lahiri (2003) is the best source for bootstrap applications to a variety of types of dependent data situations.

The model-based resampling approach to time series was first discussed by Freedman (1984) and was followed by Freedman and Peters (1984a, b), Swanepoel and van Wyk (1986), and Efron and Tibshirani (1986). Li and Maddala (1996) provided a survey of related literature on bootstrapping in the time domain with emphasis on econometric applications.

Peters and Freedman (1985) dealt with bootstrapping for the purpose of comparing competing forecasting methods. Tsay (1992) provided an applications-oriented approach to parametric bootstrapping of time series.

Kabaila (1993) discussed prediction in autoregressive time series using the bootstrap. Stoffer and Wall (1991) applied the bootstrap using state space models. Chen et al. (1993) used model-based resampling to determine the appropriate order for an autoregressive model.

The higher-order asymptotic properties for block resampling have been given by Lahiri (1991) and Gotze and Kunsch (1996). Davison and Hall (1993) provided some insight by showing that these good asymptotic properties require the right choice of a variance estimate. Lahiri (1992) applied an Edgeworth correction to the MBB for both stationary and nonstationary time series.

The DWB was introduced by Shao (2010) who also derived its asymptotic properties and showed some of the advantages it has over the various block bootstrap approaches. Another recent paper applying the DWB is McMurry and Politis (2010).

The stationary bootstrap was introduced by Politis and Romano (1994). Earlier, Politis and Romano (1992) proposed the CBB. Liu and Singh (1992) obtained general results for the use of MBB and jackknife approaches in very general types of dependent data. Other work in this area includes Liu (1988) and Liu and Singh (1995).

Fan and Hung (1997) used balanced resampling (a variance reduction technique to be discussed later) to bootstrap finite Markov chains. Liu and Tang (1996) used bootstrap for control charting in both the dependent and independent data sequences.

Frequency domain resampling has been discussed in Franke and Hardle (1992) with an analogy made to nonparametric regression. Janas (1993) and Dahlhaus and Janas (1996) extended the results of Franke and Hardle. Politis et al. (1992) provided bootstrap confidence bands for spectra and cross-spectra (Fourier transforms of autocorrelation and cross-correlation). More can be found in Davison and Hinkley (1997) and Lahiri (2003). We understand that unpublished work by Hurvich and Zeger (1987) was a key paper that preceded these references.

The sieve bootstrap was introduced and shown to be consistent by Kreiss (1992). Refinements for a class of stationary stochastic processes were given by Buhlmann (1997). See also Paparoditis and Streiberg (1992). For more details, see section 2.10 of Lahiri (2003).

5.10   EXERCISES

1. NASDAQ Closing Values: File "nasdaq.csv" contains several years by day of closing values of the NASDAQ stock index.
   (a) Log-transform the closing values (multiplicative effect model).
   (b) Use the R function arima() to fit an $MA(1)$ model after one forward different to make stationary. What are the parameter estimates and 95% confidence intervals? (Hint: arima(lnClose, order = c(0,1,1)).)
   (c) Use model-based resampling to find a 95% confidence interval for the moving average coefficient. Use $R = 1000$ resamples.
   (d) Use block resampling with $l = 20$ to find a 95% confidence interval on the moving average coefficient. Use $R = 1000$ resamples.
   (e) Compare the three confidence intervals. Which do you think is more trustworthy?

2. SP500 Closing Values: File "sp500.csv" contains several years by day of closing values of the Standard & Poors 500 stock index.
   (a) Log-transform the closing values (multiplicative effect model).
   (b) Use the R function arima() to fit an $MA(1)$ model after one forward different to make stationary. What are the parameter estimates and 95% confidence intervals? (Hint: arima(lnClose, order = c(0,1,1)).)
   (c) Use model-based resampling to find a 95% confidence interval for the moving average coefficient. Use $R = 1000$ resamples.
   (d) Use block resampling with $l = 20$ to find a 95% confidence interval on the moving average coefficient. Use $R = 1000$ resamples.
   (e) Compare the three confidence intervals. Which do you think is more trustworthy?

REFERENCES

Anderson, T. W. (1971). The Statistical Analysis of Time Series. Wiley, New York.

Basawa, I. V., Mallik, A. K., McCormick, W. P., and Taylor, R. L. (1989). Bootstrapping explosive first order autoregressive processes. Ann. Stat. 17, 1479–1486.

Basawa, I. V., Mallik, A. K., McCormick, W. P., Reeves, J. H., and Taylor, R. L. (1991a). Bootstrapping unstable first order autoregressive processes. Ann. Stat. 19, 1098–1101.

Basawa, I. V., Mallik, A. K., McCormick, W. P., Reeves, J. H., and Taylor, R. L. (1991b). Bootstrapping tests of significance and sequential bootstrap estimators for unstable first order autoregressive processes. Commun. Stat. Theory Methods 20, 1015–1026.

Bloomfield, P. (1976). Fourier Analysis of Time Series: An Introduction. Wiley, New York.

Bollerslev, T. (1986). Generalized autoregressive conditional heteroskedasticity. J. Econom. 31, 307–327.

Bose, A. (1988). Edgeworth correction by bootstrap in autoregressions. Ann. Stat. 16, 1709–1726.

Box, G. E. P., and Jenkins, G. M. (1970). Time Series Analysis: Forecasting and Control. Holden Day, San Francisco, CA.

Box, G. E. P., and Jenkins, G. M. (1976). Time Series Analysis: Forecasting and Control, 2nd Ed. Holden Day, San Francisco, CA.

Box, G. E. P., Jenkins, G. M., and Reinsel, G. C. (1994). Time Series Analysis: Forecasting and Control, 3rd Ed. Prentice-Hall, Englewood Cliffs, NJ.

Braun, W. J., and Kulperger, P. J. (1997). Properties of a Fourier bootstrap method for time series. Commun. Stat. Theory Methods 26, 1326–1336.

Brillinger, D. R. (1981). Time Series: Data Analysis and Theory, Expanded Edition. Holt, Reinhart, and Winston, New York.

Brockwell, P. J., and Davis, R. A. (1991). Time Series Methods, 2nd Ed. Springer-Verlag, New York.

Buhlmann, P. (1997). Sieve bootstrap for time series. Bernoulli 3, 123–148.

Buhlmann, P. (2002). Sieve bootstrap with variable-length Markov chains for stationary categorical time series. J. Am. Stat. Assoc. 97, 443–456.

Carlstein, E. (1986). The use of subseries values for estimating the variance of a general statistic from a stationary sequence. Ann. Stat. 14, 1171–1194.

Chatterjee, S. (1986). Bootstrapping ARMA models. Some simulations. IEEE Trans. Syst. Man. Cybern. 16, 294–299.

Chen, C., Davis, R. A., Brockwell, P. J., and Bai, Z. D. (1993). Order determination for autoregressive processes using resampling methods. Stat. Sin. 3, 481–500.

Chernick, M. R. (2007). Bootstrap Methods: A Guide for Practitioners and Researchers, 2nd Ed. Wiley, Hoboken, NJ.

Chernick, M. R., Downing, D. J., and Pike, D. H. (1982). Detecting outliers in time series data. J. Am. Stat. Assoc. 77, 743–747.

Chernick, M. R., Daley, D. J., and Littlejohn, R. P. (1988). A time-reversibility relationship between two Markov chains with exponential stationary distributions. J. Appl. Probab. 25, 418–422.

Dahlhaus, R., and Janas, D. (1996). A frequency domain bootstrap for ratio statistics in time series analysis. Ann. Stat. 24, 1914–1933.

Datta, S. (1995). Limit theory and bootstrap for explosive and partially explosive autoregression. Stoch. Proc. Appl. 57, 285–304.

Datta, S. (1996). On asymptotic properties of bootstrap for AR (1) processes. J. Stat. Plan. Inference 53, 361–374.

Datta, S., and Sriram, T. N. (1997). A modified bootstrap for autoregressions without stationarity. J. Stat. Plan. Inference 59, 19–30.

Davison, A. C., and Hall, P. (1993). On studentizing and blocking methods for implementing the bootstrap with dependent data. Aust. J. Statist. 35, 215–224.

Davison, A. C., and Hinkley, D. V. (1997). Bootstrap Methods and Their Applications. Cambridge University Press, Cambridge.

Efron, B. (1982). The Jackknife, the Bootstrap and Other Resampling Plans. SIAM, Philadelphia.

Efron, B., and Tibshirani, R. (1986). Bootstrap methods for standard errors, confidence intervals and other measures of statistical accuracy. Stat. Sci. 1, 54–77.

Engel, R. (1982). Autoregressive conditional heteroscedasticity with estimates of the variance of UK inflation. Econometrica 50, 987–1008.

Fan, J., and Koul, H. L. (eds.). (2006). Frontiers in Statistics. Imperial College Press, London.

Fan, T.-H., and Hung, W.-L. (1997). Balanced resampling for bootstrapping finite Markov chains. Commun. Stat. Simul. Comput. 26, 1465–1475.

Findley, D. F. (1986). On bootstrap estimates of forecast mean square error for autoregressive processes. Proc. Comput. Sci. Stat. 17, 11–17.

Franke, J., and Hardle, W. (1992). On bootstrapping kernel spectral estimates. Ann. Stat. 20, 121–145.

Freedman, D. A. (1984). On bootstrapping two-stage least squares estimates in stationary linear models. Ann. Stat. 12, 827–842.

Freedman, D. A., and Peters, S. C. (1984a). Bootstrapping a regression equation: Some empirical results. J. Am. Stat. Assoc. 79, 97–106.

Freedman, D. A., and Peters, S. C. (1984b). Bootstrapping an econometric model: Some empirical results. J. Bus. Econ. Stat. 2, 150–158.

Fuller, W. A. (1976). Introduction to Statistical Time Series. Wiley, New York.

Gotze, F., and Kunsch, H. R. (1996). Second-order correctness of the blockwise bootstrap for stationary observations. Ann. Stat. 24, 1914–1933.

Hall, P., and Jing, B. Y. (1996). On sample reuse methods for dependent data. J. R. Stat. Soc. B 58, 727–737.

Hamilton, J. D. (1994). Time Series Analysis. Princeton University Press, Princeton, NJ.

Harvey, A. C., Ruiz, E., and Shepard, N. (1994). Multivariate stochastic variance models. Rev. Econ. Stud. 61, 247–264.

Holbert, D., and Son, M.-S. (1986). Bootstrapping a time series model: Some empirical results. Commun. Stat. Theory Methods 15, 3669–3691.

Hurvich, C. M., and Zeger, S. L. (1987). Frequency domain bootstrap methods for time series. Statistics and Operations Research Working Paper, New York University, New York.

Janas, D. (1993). Bootstrap Procedures for Time Series. Verlag Shaker, Aachen.

Kabaila, P. (1993). On bootstrap predictive inference for autoregressive processes. J. Time Ser. Anal. 14, 473–484.

Kreiss, J. P. (1992). Bootstrap procedures for AR(∞) processes. In Bootstrapping and Related Techniques, Proceeding, Trier, FRG. Lecture Notes in Economics and Mathematical Systems (K.-H. Jockel, G. Rothe, and W. Sendler, eds.), pp. 107–113. Springer-Verlag, Berlin.

Kreiss, J. P., and Franke, J. (1992). Bootstrapping stationary autoregressive moving average models. J. Time Ser. Anal. 13, 297–317.

Kunsch, H. (1989). The jackknife and bootstrap for general stationary observations. Ann. Stat. 17, 1217–1241.

Lahiri, S. N. (1991). Second order optimality of stationary bootstrap. Stat. Probab. Lett. 14, 335–341.

Lahiri, S. N. (1992). Edgeworth correction by moving block bootstrap for stationary and non-stationary data. In Exploring the Limits of Bootstrap (R. LaPage, and L. Billard, eds.), pp. 183–214. Wiley, New York.

Lahiri, S. N. (2002). Comment on "Sieve bootstrap with variable-length Markov chains for stationary categorical time series". J. Am. Stat. Assoc. 97, 460–461.

Lahiri, S. N. (2003). Resampling Methods for Dependent Data. Springer-Verlag, New York.

Li, H., and Maddala, G. S. (1996). Bootstrapping time series models (with discussion). Econ. Rev. 15, 115–195.

Liu, R. Y. (1988). Bootstrap procedures under some non-i.i.d. models. Ann. Stat. 16, 1696–1708.

Liu, R. Y., and Singh, K. (1992). Moving blocks jackknife and bootstrap capture weak dependence. In Exploring the Limits of Bootstrap (R. LaPage, and L. Billard, eds.), pp. 225–248. Wiley, New York.

Liu, R. Y., and Singh, K. (1995). Using i.i.d. bootstrap for general non-i.i.d. models. J. Stat. Plan. Inference 43, 67–76.

Liu, R. Y., and Tang, J. (1996). Control charts for dependent and independent measurements based on bootstrap methods. J. Am. Stat. Assoc. 91, 1694–1700.

Martin, R. D. (1980). Robust estimation of autoregressive models. In Reports on Directions in Time Series (D. R. Brillinger, and G. C. Tiao, eds.), pp. 228–262. Institute of Mathematical Statistics, Hayward, CA.

McCullough, B. D. (1994). Bootstrapping forecast intervals: An application to AR(p) models. J. Forecast. 13, 51–66.

McMurry, T., and Politis, D. N. (2010). Banded and tapered estimates of autocovariance matrices and the linear process bootstrap. J. Time Ser. Anal. 31, 471–482.

Melino, A., and Turnbull, S. M. (1990). Pricing foreign currency options with stochastic volatility. J. Econ. 45, 239–265.

Nelson, D. B. (1991). Conditional heteroskedasticity in asset returns: A new approach. Econometrica 59, 347–370.

Nicholls, D. F., and Quinn, B. G. (1982). Random Coefficient Autoregressive Models: An Introduction. Springer-Verlag, New York.

Paparoditis, E., and Politis, D. N. (2001). Tapered block bootstrap. Biometrika 88, 1105–1119.

Paparoditis, E., and Politis, D. N. (2002). The tapered block bootstrap for general statistics from stationary sequences. Econom. J. 5, 131–148.

Paparoditis, E., and Streiberg, B. (1992). Order identification statistics in stationary autoregressive moving-average models: Vector autocorrelations and the bootstrap. J. Time Ser. Anal. 13, 415–434.

Pena, D., Tiao, G. C., and Tsay, R. S. (eds.). (2001). A Course in Time Series Analysis. Wiley, New York.

Peters, S. C., and Freedman, D. A. (1985). Using the bootstrap to evaluate forecasts. J. Forecast. 4, 251–262.

Politis, D. N., and Romano, J. P. (1992). A circular block-resampling procedure for stationary data. In Exploring the Limits of Bootstrap (R. LaPage, and L. Billard, eds.), pp. 263–270. Wiley, New York.

Politis, D. N., and Romano, J. P. (1994). The stationary bootstrap. J. Am. Stat. Assoc. 89, 1303–1313.

Politis, D. N., Romano, J. P., and Lai, T. L. (1992). Bootstrap confidence bands for spectra and cross-spectra. IEEE Trans. Signal Process. 40, 1206–1215.

Priestley, M. B. (1981). Spectral Analysis of Time Series. Academic Press, London.

Rousseeuw, P. J., and Leroy, A. M. (1987). Robust Regression and Outlier Detection. Wiley, New York.

Shao, X. (2010). The dependent wild bootstrap. J. Am. Stat. Assoc. 105, 218–235.

Shao, Q., and Yu, H. (1993). Bootstrapping the sample means for stationary mixing sequences. Stoch. Proc. Appl. 48, 175–190.

Stine, R. A. (1987). Estimating properties of autoregressive forecasts. J. Am. Stat. Assoc. 82, 1072–1078.

Stoffer, D. S., and Wall, K. D. (1991). Bootstrapping state-space models: Gaussian maximum likelihood estimation and Kalman filter. J. Am. Statist. Assoc. 86, 1024–1033.

Swanepoel, J. W. H., and van Wyk, J. W. J. (1986). The bootstrap applied to power spectral density function estimation. Biometrika 73, 135–141.

Thombs, L. A. (1987). Recent advances in forecasting: Bootstrap prediction intervals for time series. In 1987 Proceedings of the Decision Sciences Institute (Bob Parsons and John Saber editors).

Thombs, L. A., and Schucany, W. R. (1990). Bootstrap prediction intervals for autoregression. J. Am. Stat. Assoc. 85, 486–492.

Tong, H. (1983). Threshold Models in Non-Linear Time Series Analysis. Springer-Verlag, New York.

Tong, H. (1990). Non-Linear Time Series: A Dynamical Systems Approach. Clarendon Press, Oxford, UK.

Tsay, R. S. (1987). Conditional heteroskedastic time series models. J. Am. Stat. Assoc. 82, 590–604.

Tsay, R. S. (1992). Model checking via parametric bootstraps in time series. Appl. Stat. 41, 1–15.

Weigend, A. S., and Gershenfeld, N. A. (1994). Time Series Prediction: Forecasting the Future and Understanding the Past. Addison-Wesley Publishing Company, New York.

Yule, G. U. (1927). On a method for investigating periodicities in disturbed series, with special reference to the Wolfer sunspot data. Philos. Trans. A 226, 267–298.

6

BOOTSTRAP VARIANTS

In previous chapters, we have introduced some modifications to the nonparametric bootstrap. These modifications were sometimes found to provide improvements especially when the sample size is small. For example, in the error rate estimation problem for linear discriminant functions the 632 estimator and double bootstrap were introduced among others.

In the case of confidence intervals, Hall has shown that the accuracy of both types of percentile method bootstrap confidence intervals can be improved on through bootstrap iteration. If we expect that the data come from an absolutely continuous distribution, kernel methods may be used to smooth the empirical distribution function and have the bootstrap samples taken from the smoothed distribution. Consequently, such an approach is called a smoothed bootstrap.

Even though it is desirable to smooth the empirical distribution in small samples, there is a catch because kernel methods need large samples to accurately estimate the tails of the distribution. There is also a variance/bias trade-off associated with the degree of smoothing. In many applications of kernel methods, cross-validation is used to determine what the smoothing parameter should be. So, in general, it is not clear when smoothing will actually help. Silverman and Young (1987) gave a great deal of attention to this issue.

An Introduction to Bootstrap Methods with Applications to R, First Edition. Michael R. Chernick, Robert A. LaBudde.
© 2011 John Wiley & Sons, Inc. Published 2011 by John Wiley & Sons, Inc.

Sometimes in density estimation or spectral density estimation, the bootstrap itself may be looked at as a technique to decide on what the smoothing parameter should be. Books on density estimation such as Silverman (1986) indicate that for kernel methods, the shape of the kernel is not a major issue but the degree of smoothing is.

Another variant to the bootstrap is the Bayesian bootstrap that was introduced in Rubin (1981). The Bayesian bootstrap can be viewed as a Bayesian justification for using bootstrap methods. This is how Efron and some other bootstrap advocates view it. On the other hand, Rubin used it to point out some weakness in the frequentist bootstrap. Inspite of that, there are now a number of interesting applications of the Bayesian bootstrap, particularly in the context of missing data (see, e.g., Lavori et al., 1995 and chapter 12 in Dmitrienko et al., 2007). In the next section, we cover the Bayesian bootstrap in some detail.

6.1 BAYESIAN BOOTSTRAP

Consider the case where $x_1, x_2, \ldots, x_n$ can be viewed as a sample of n-independent identically distributed realizations of the random variables $X_1, X_2, \ldots, X_n$ each with distribution F, and denote the empirical distribution by $\hat{F}$. The nonparametric bootstrap samples from $\hat{F}$ with replacement. Let θ be a parameter of the distribution F. For simplicity, we may think of x_1 as one-dimensional and θ is a single parameter but they both could be vectors. Let $\hat{\theta}$ be an estimate of θ based on $x_1, x_2, \ldots, x_n$. We know that the nonparametric bootstrap can be used to approximate the distribution of $\hat{\theta}$.

Instead of sampling each x_i independently with replacement and equal probability $1/n$, the Bayesian bootstrap uses a posterior probability distribution for the X_i's. This posterior probability distribution is centered at $1/n$ for each X_i, but the probability of selection changes from sample to sample. Think of it as an n-dimensional vector with mean putting equal weight $1/n$ on each X_i but the actual proportion for each X_i on the kth draw is drawn from the posterior distribution. Specifically, the Bayesian bootstrap replications are defined as follows. Draw $n - 1$ uniform random variables on the interval $[0, 1]$, denote the ordered values from minimum to maximum as $u_{(1)}, u_{(2)}, \ldots, u_{(n-1)}$, and define $u_{(0)} = 0$ and $u_{(n)} = 1$. Define $g_i = u_{(i)} - u_{(i-1)}$ for $i = 1, 2, \ldots, n$, defined in this way:

$$\sum_{i=1}^{n} g_i = 1.$$

The g_i's are called the uniform spacings or gaps. Their distribution theory can be found in David (1981). Now, suppose we take n Bayesian bootstrap replications and let $g_i^{(k)}$ be the probability associated with X_i on the kth Bayesian bootstrap replication.

From David (1981), we have $E\left(g_i^{(k)}\right) = 1/n$ for each i and k, $var\left(g_i^{(k)}\right) = (n-1)/n^3$, and $cov\left(g_i^{(k)}, g_j^{(k)}\right) = -1/(n-1)$ for each k. Because of these properties, the bootstrap distribution for $\hat{\theta}$ is very similar to the Bayesian bootstrap distribution for θ in many applications. Rubin (1981) provided some examples and showed that the Bayesian bootstrap procedure leads to a posterior distribution, which is Dirichlet and is based on a conjugate prior for the Dirichlet.

Because the prior distribution appears to be unusual, Rubin thought the examples show reason to question the use of the Bayesian bootstrap. In his view, it may be appropriate in some instances but not in general. The main point is that if there is a reason to question the Bayesian bootstrap because of the prior, by analogy, the ordinary nonparametric bootstrap is so similar that it too should be questioned. The criticism is primarily based on the lack of smoothness of the empirical distribution. So smoothing the bootstrap should get around that problem.

The Bayesian bootstrap has been generalized to allow for posterior distributions that are not Dirichlet. A good reference for the developments over 15 years following Rubin's paper from 1981 is Rubin and Schenker (1998).

6.2 SMOOTHED BOOTSTRAP

One reason that influences the choice of resampling from the empirical distribution function is that it represents the maximum likelihood estimate for any of the distributional parameters under minimal assumptions (i.e., the nonparametric mle). However, in many applications, it is reasonable to assume that the distribution is absolutely continuous. In that case, it would make sense to take advantage of this information to compute a density estimate or "smoothed version" of the estimate of F. One way to do this is through kernel smoothing methods. A Bayesian approach to this that we will not discuss here is given in Banks (1988).

Efron (1982) illustrated the use of a smoothed version of the bootstrap for the correlation coefficient using both Gaussian and uniform kernels. In comparing the bootstrap estimate by simulation when the observations are taken to be Gaussian with the nonparametric (unsmoothed) version, he finds some improvement in the estimate of the standard error for the correlation coefficient estimate.

Although early on, in the development of the bootstrap, smoothed versions of F were considered; Dudewicz and others have only more recently given it consideration and done extensive comparisons with the unsmoothed version. Dudewicz (1992) called it a generalized bootstrap because it allows for a wider choice of estimates without relaxing the distributional assumptions very much. Rather than use kernel methods to estimate the density, Dudewicz suggested using a very broad class of distributions to fit to the observed data. The fitted distribution is then used to draw bootstrap samples.

One such family is called the generalized lambda distribution (see Dudewicz, 1992, p. 135). The generalized lambda distribution is a four-parameter family that can be specified by its mean, variance, skewness, and kurtosis. The method of moment estimates of these four parameters is recommended for fitting this distribution. For a particular application, Sun and Muller-Schwarze (1996) compared the generalized bootstrap obtained as just described with the nonparametric bootstrap.

It appears that the generalized bootstrap might be an improvement over the kernel smoothing approach as it does not make very restrictive parametric assumptions but still provides a continuous distribution without the usual issue that come with kernel smoothing methods. Another approach to obtain a generalized bootstrap is presented in Bedrick and Hill (1992).

6.3 PARAMETRIC BOOTSTRAP

Efron (1982) called the nonparametric bootstrap a nonparametric maximum likelihood estimation procedure. In this way, it can be viewed as an extension of Fisher's maximum likelihood approach. However, if we further assume that the distribution for the samples F is absolutely continuous, then it is more appropriate to smooth F to incorporate this additional knowledge. We could take this a step further and assume that F comes from a parametric family. In that case, Fisher's theory of maximum likelihood could be applicable as are other methods for estimating a small number of parameters.

So if an estimate of F is chosen from a parametric family, then sampling with replacement from that distribution could be considered a parametric bootstrap. The parametric bootstrap is discussed briefly on pages 29–30 of Efron (1982). If maximum likelihood is used to estimate the parameters of the distribution F, then the approach is essentially the same as maximum likelihood. So bootstrapping usually does not add anything to parametric problems. In complex problems, however, it may be helpful to have at least partial parameterization. These semiparametric methods (e.g., the Cox proportional hazards model) can be handled by the bootstrap and sometimes are put under the category of the parametric boostrap.

Davison and Hinkley (1997) provided other justification for a parametric bootstrap. First of all a comparison of the nonparametric bootstrap with a parametric bootstrap provides a check on the parametric assumptions. They used the exponential distribution to illustrate the parametric bootstrap and pointed out that it is useful when the parametric distribution is difficult to derive or has an asymptotic approximation that is not accurate in small sample sizes.

6.4 DOUBLE BOOTSTRAP

The double bootstrap is a special case of bootstrap iteration where the number of iterations is 1. Because it is so highly computer-intensive it is often avoided. However, sometimes the advantages are large enough that the gain from using only one iteration is sufficient. For example, Efron (1983) used a double bootstrap as one of the methods to adjust bias of the apparent error rate in the classification problem. It was also considered by other authors in subsequent simulation studies.

Using the brute force approach to the double bootstrap would require generating B^2 bootstrap samples if B is both the number of bootstrap samples taken from the original sample and the number of bootstrap samples taken for each adjustment to the estimate from an original bootstrap sample. More generally, if the number of initial bootstrap samples is B_1 and each sample is adjusted using B_2 bootstrap samples the total number of bootstrap samples that would be generated is $B_1 B_2$. But in Efron (1983), Efron provided a variance reduction technique that allows the same accuracy as in the brute force method where B^2 bootstrap samples are needed but only $2B$ bootstrap samples are actually used.

Bootstrap iteration has been shown to improve the order of accuracy of bootstrap confidence intervals and the theory is due to Hall, Beran, Martin, and perhaps a few others. This theory is nicely summarized in Hall (1992, chapter 3, section 3.1.4). But we must keep in mind that the increased accuracy comes with a price, better results with increased computation time.

6.5   THE *M*-OUT-OF-*N* BOOTSTRAP

When the nonparametric bootstrap was first introduced, Efron proposed taking the sample size for the bootstrap sample to be the same as the sample size n from the original sample. This seemed reasonable and worked quite well in many applications. One might argue that since the idea of the bootstrap sample is to mimic the behavior of the original sample and the accuracy of sample estimates generally depends on the sample size n, choosing a sample size m larger than n would lead to an estimate with less variability than the original sample, while a sample size m smaller than n would cause the estimator to have more variability.

Nevertheless, some early on suggested that it might be reasonable to look at $m < n$. This might make sense when sampling from a finite population or for other dependencies such as time series or spatial data. The argument might be that sampling independently with replacement in the dependent situations creates estimates with variability less than what we actually have in the original data set since n-dependent observations contain only the information equivalent to a smaller number of independent observations. So if bootstrapping behaves with weaker dependence than the original sample, a smaller number of observations would mimic the variability of the estimate better.

In fact, in Chapter 8, we will see that an m-out-of-n bootstrap with $m \ll n$ will cure inconsistency problems associated with survey sampling and other dependent cases. But perhaps surprisingly, it also works in independent cases such as when the mean converges to a stable law other than the normal distribution or when estimating the asymptotic behavior of the maximum of an independent and identically distributed (i.i.d.) sequence of random variables. This method has been studied by many authors, and asymptotic theory has been developed to show that if $m \to \infty$ and $n \to \infty$ at a rate such that $m/n \to 0$, the m-out-of-n bootstrap is consistent in situations where n-out-of-n would not be. For a detailed account of this theory, see Bickel et al. (1997).

In practice, the m-out-of-n bootstrap would only be used in situations where the ordinary bootstrap is known or suspected to be inconsistent. Otherwise, confidence intervals would be wider. There is really no reason not to use $m = n$ in situations where the bootstrap is consistent.

6.6 THE WILD BOOTSTRAP

We describe the wild bootstrap and its use in linear regression with heteroscedastic error terms. Here we provide the formal definition and mention some variants of the original proposal of Liu (1988). Consider the linear model $Y = X\beta + u$ and, according to this model, observed values $y_i = X_i\beta + u_i$, where the u_is are the error terms independent of each other, the X_i are the values of the k covariates, and β is a vector of k parameters corresponding to their respective covariates. Let $\hat{\beta}$ be the ordinary least squares estimate of β and $\hat{u}_i$ be the least squares residuals. Bootstrap residuals then go ahead and create a bootstrap sample from the residuals, use the model with the least squares estimates of β to generate the bootstrap sample of observations, and then use the least squares procedure on the bootstrap observations to obtain a bootstrap estimate of β. This is repeated B times to generate an approximation to the bootstrap distribution of $\hat{\beta}$. From this bootstrap distribution, the estimated covariance matrix for the parameters can be obtained from which we get estimates of standard errors for the regression parameters. This also could be done using a different estimation procedure such as the least absolute errors. This was covered in greater generality in Chapter 2.

The wild bootstrap does the same thing except that the least squares residuals are modified and the modified residuals are bootstrapped. Specifically, the wild bootstrap residuals are obtained as follows: $u_i^* = h_i(\hat{u}_i)\varepsilon_i$, where ε_i has mean 0 and variance 1, and h_i is a transformation.

In the heteroscedastic model, authors have suggested taking $h_i(\hat{u}_i) = \hat{u}_i/(1-H_i)^{1/2}$ or $\hat{u}_i/(1-H_i)$, where $H_i = X_i(X^T X)^{-1} X_i^T$, the ith diagonal element of orthogonal projection matrix (called the hat matrix in the statistics literature). So the variants of the wild bootstrap come about by applying different choices for h_i and the distribution of the ε_i. Here, we have seen two choices for h_i and in Chapter 2 two choices for the distribution of ε_i. Also, some authors restrict the distribution to have $E(\varepsilon_i^3) = 1$.

6.7 HISTORICAL NOTES

The Bayesian bootstrap was introduced in Rubin (1981). A good source covering the literature on it up to 1998 can be found in Rubin and Schenker (1998). As with many of the concepts, the smoothed bootstrap was discussed early on in Efron (1982). Silverman and Young (1987) provided the best examples of difficulties and advantages of the technique. Dudewicz (1992) called it the generalized bootstrap and used it in a parametric form that covers a broad class of distributions. This parametric approach was also first discussed in Efron (1982), but the best advocates for it are Davison and Hinkley (1997). The double (or iterated) bootstrap was first used on an applied problem in Efron (1983). The work of Hall (1992) is a great source to guide the reader to the theory of it along with many references.

Early examples where m-out-of-n bootstrapping was proposed include Bretagnolle (1983), Swanepoel (1986), Athreya (1987), and Politis and Romano (1993). Bickel and Ren (1996) showed promise in the m-out-of-n bootstrap and brought it to the spotlight. But its many virtues came out when it was found to be a way to obtain consistent estimates. Bickel et al. (1997) gave a very detailed account of the pros and cons for m-out-of-n.

The wild bootstrap came about from the discussion by Beran and Wu in Wu (1986). The work of Liu (1988) was about the first formal treatment of the wild bootstrap. Mammen (1993) used a modification of the distribution for ε_i and showed its good asymptotic properties.

Davidson and Flachaire (2008) constructed a bootstrap test based on the wild bootstrap where additional modifications are employed to improve asymptotic results. Flachaire (2003) showed by simulation that the wild bootstrap using Liu's formulation of the ε_i's is superior to Mammen's and does better than bootstrapping by pairs. A very recent paper that also looks at the dependent wild bootstrap is McMurry and Politis (2010).

Heteroscedastic regression models appear frequently in the econometrics literature, and hence, many of the papers on the wild bootstrap do also. Examples include Davidson and MacKinnon (1985, 1998, 1999), Flachaire (1999), Horowitz (2000), MacKinnon (2002), and van Giersbergen and Kiviet (2002).

6.8 EXERCISE

1. Surimi data: Using the surimi data of Example 3.1, estimate the Efron percentile 95% confidence interval for the mean using an "m-of-n" bootstrap with $m = 20$.
2. Compare with the 95% confidence interval obtained using all the data.

REFERENCES

Athreya, K. B. (1987). Bootstrap estimate of the mean in the infinite variance case. Ann. Stat. 15, 724–731.

Banks, D. L. (1988). Histospline smoothing the Bayesian bootstrap. Biometrika 75, 673–684.

Bedrick, E. J., and Hill, J. R. (1992). A generalized bootstrap. In Exploring the Limits of Bootstrap (R. LePage, and L. Billard, eds.). Wiley, New York.

Bickel, P. J., and Ren, J. J. (1996). The m out of n bootstrap and goodness of fit tests for doubly censored data. In Robust Statistics, Data Analysis, and Computer-Intensive Methods (P. J. Huber, and H. Rieder, eds.), pp. 35–48. Springer-Verlag, New York.

Bickel, P. J., Gotze, F., and van Zwet, W. R. (1997). Resampling fewer than n observations, gains, losses, and remedies for losses. Stat. Sin. 7, 1–32.

Bretagnolle, J. (1983). Lois limites du bootstrap de certaines fonctionelles. Ann. Inst. Henri Poincare Sect. B 19, 281–296.

David, H. A. (1981). Order Statistics. Wiley, New York.

Davidson, R., and Flachaire, E. (2008). The wild bootstrap, tamed at last. J. Econom. 146, 162–169.

Davidson, R., and MacKinnon, J. G. (1985). Heteroskedasticity-robust tests in regression directions. Annales l'INSEE 59/60, 183–218.

Davidson, R., and MacKinnon, J. G. (1998). Graphical methods for investigating the size and power of hypothesis tests. Manchester Sch. 66, 1–26.

Davidson, R., and MacKinnon, J. G. (1999). The size and distortion of bootstrap tests. Econom. Theory 15, 361–376.

Davison, A. C., and Hinkley, D. V. (1997). Bootstrap Methods and Their Applications. Cambridge University Press, Cambridge.

Dmitrienko, A., Chuang-Stein, C., and D'Agostino, R. (eds.). (2007). Pharmaceutical Statistics Using SAS? A Practical Guide. SAS Institute, Inc., Cary, NC.

Dudewicz, E. J. (1992). The generalized bootstrap. In Bootstrapping and Related Techniques, Proceedings, Tier, FRG, Lecture Notes in Economics and Mathematical Systems 376 (K.-H. Jockel, G. Rothe, and W. Sendler, eds.), pp. 31–37. Springer-Verlag, Berlin.

Efron, B. (1982). The Jackknife, the Bootstrap, and Other Resampling Plans. SIAM, Philadelphia.

Efron, B. (1983). Estimating the error rate of a prediction rule: Improvements on cross-validation. J. Am. Stat. Assoc. 78, 316–331.

Flachaire, E. (1999). A better way to bootstrap pairs. Econom. Lett. 64, 257–262.

Flachaire, E. (2003). Bootstrapping heteroskedastic regression models: Wild bootstrap versus pairs bootstrap. Technical report EUREQUA, Univesite Paris 1 Pantheon-Sorbomme.

Hall, P. (1992). The Bootstrap and Edgeworth Expansion. Springer-Verlag, New York.

Horowitz, J. L. (2000). The bootstrap. In Handbook of Econometrics 5 (J. J. Heckman, and E. E. Leamer, eds.). Elsevier Science, Holland.

Lavori, P. W., Dawson, R., and Shera, D. (1995). A multiple imputation strategy for clinical trials with truncation of patient data. Stat. Med. 14, 1913–1925.

Liu, R. Y. (1988). Bootstrap procedures under some non i.i.d. models. Ann. Stat. 16, 1696–1708.

MacKinnon, J. G. (2002). Bootstrap inference in econometrics. Can. J. Econ. 35, 615–645.

Mammen, E. (1993). Bootstrap and wild bootstrap for high dimensional linear models. Ann. Stat. 21, 255–285.

McMurry, T., and Politis, D. N. (2010). Banded and tapered estimates of autocovariance matrices and the linear process bootstrap. J. Time Ser. Anal. 31, 471–482.

Politis, D. N., and Romano, J. P. (1993). Estimating the distribution of a studentized statistic by subsampling. Bull. Intern. Stat. Inst. 49th Session 2, 315–316.

Rubin, D. B. (1981). The Bayesian bootstrap. Ann. Stat. 9, 130–134.

Rubin, D. B., and Schenker, N. (1998). Imputation. In Encyclopedia of Statistical Sciences, Vol. 2 (S. Kotz, C. B. Read, and D. L. Banks, eds.). Wiley, New York.

Silverman, B. W. (1986). Density Estimation for Statistics and Data Analysis. Chapman & Hall, London.

Silverman, B. W., and Young, G. A. (1987). The bootstrap. To smooth or not to smooth? Biometrika 74, 469–479.

Sun, L., and Muller-Schwarze, D. (1996). Statistical resampling methods in biology: A case study of beaver dispersal patterns. Am. J. Math. Manag. Sci. 16, 463–502.

Swanepoel, J. W. H. (1986). A note on proving that the (modified) bootstrap works. Commun. Stat. Theory Methods 12, 2059–2083.

van Giersbergen, N. P. A., and Kiviet, J. F. (2002). How to implement bootstrap hypothesis testing in static and dynamic regression models: Test statistic versus confidence interval approach. J. Econom. 108, 133–156.

Wu, C. F. J. (1986). Jackknife, bootstrap and other resampling plans in regression analysis (with discussion). Ann. Stat. 14, 1261–1350.

7

CHAPTER SPECIAL TOPICS

This chapter deals with a variety of statistical problems that can be effectively handled by the bootstrap. The only common theme is that the problems are complex. In many cases, classical approaches require special (restrictive) assumptions or provide incomplete or inadequate answers.

Although the bootstrap works well in these cases, theoretical justification is not always available, and just like with the error rate estimation problem, we rely on simulation studies. Efron and others felt that, although the bootstrap works well in simple problems and can be justified theoretically, it would be with the complex and intractable problems where it would have its biggest payoff.

7.1 SPATIAL DATA

7.1.1 Kriging

There are many applications of kriging, which is a technique that takes data at certain grid points in a region and generates a smooth two-dimensional curve to estimate the level of the measured variable at places between the observed values. So it is a form

An Introduction to Bootstrap Methods with Applications to R, First Edition. Michael R. Chernick, Robert A. LaBudde.
© 2011 John Wiley & Sons, Inc. Published 2011 by John Wiley & Sons, Inc.

of spatial interpolation and is the most common way to fit such surfaces. In many applications, the measurements are taken over time. Sometimes the temporal element is important, and then spatial–temporal models are sought. We will not deal with those models in this text. Kriging is a procedure done at a fixed point in time. The surface can be generated by the method at several points in time.

Graphically, the surface is pictured as contours of constant level for the estimated surface. A video could be made to show how these curves evolve over time and that may be useful, but it does not take into account the time dependence that a spatial–temporal model would and hence has limited value.

One common example is the air pollution monitoring stations that monitor the levels of particulates in the air. At any given time point, measurements are taken and can be used to generate a surface of the levels of that particulate in the region. A weather map for temperature or barometric pressure can also be generated by kriging. Missing from the map, however, is the variability due to the kriging estimation technique and the measurement. There are statistical measures that characterize this uncertainty (e.g., the variogram). However, these methods do not give a visual way to see the variability in the contours. But, if we can bootstrap the spatial data and generate bootstrap kriging contours, we could visually compare the maps generated by the bootstrap samples.

This approach was used by Switzer and Eynon as part of the research grant that SIAM awarded to Stanford for the study of air pollution. The results are presented by Diaconis and Efron (1983) in *Scientific American* as one example illustrating the potential of the bootstrap in complex problems.

The problem at the time involved the controversy of the increase in acid rain as a result of industrial pollution in the northeastern United States. They had available 2000 measurements of pH levels measured from rain accumulated at nine weather stations across the northeast over a period from September 1978 to August 1980. We will not go into the details of the kriging method here. The interested reader should consult Cressie (1991) or Stein (1999) for details about kriging.

Our focus here is on the maps that are produced and not on the particulars of the estimation method. Of course, for each bootstrap sample, the kriging solution must be computed, and this makes the computations somewhat intensive if a large number of bootstrap samples are to be generated. However, in the example, only five bootstrap maps were generated and a visual comparison of the five along with the original contours was generated. See Figure 7.1 taken from Diaconis and Efron (1983, pp. 117) with permission from *Scientific American*.

The weather stations are marked on the maps as dots. There are nine stations. They are labeled on the original map and just represented by the corresponding dots in the bootstrap maps.

The stations are (1) Montague in Massachusetts, (2) Duncan Falls in Ohio, (3) Roanoke in Indiana, (4) Rockport in Indiana, (5) Giles County in Tennessee, (6) Raleigh in North Carolina, (7) Lewisburg in West Virginia, (8) Indian River in Delaware, and (9) Scranton in Pennsylvania. It appears that in the original map, the contour 4.5 running through New York and Pennsylvania is quite different from any of the bootstrap samples where the similar contour is 4.2.

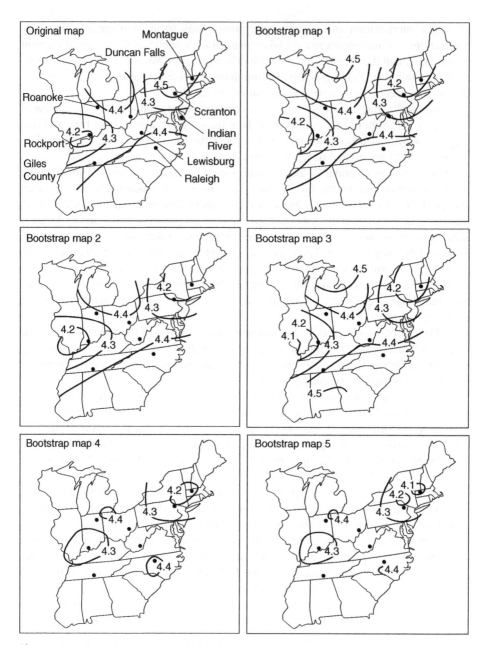

Figure 7.1. The original map and five bootstrap maps. (From Diaconis and Efron [1983, p. 117], taken with permission from *Scientific American*.)

The closest weather stations that would have the greatest influence are Scranton and Montague. Because Scranton is in an industrial area, a high contour might not be a surprise as pollution from New York, New Jersey, and Philadelphia could affect the measurements. It would seem that because of the nine stations being sampled with replacement, a high measurement at Scranton would not have an effect on the contours if it is not included in the bootstrap sample. Also, the stations that appear twice or more heavily affect the nearby contour level estimates.

Either the number 4.5 is a mistake in the original figure or Scranton is missing from all five bootstrap samples. Aside from that one number, the original and bootstrap maps 1, 2, and 3 look very similar, but markedly different from maps 4 and 5.

Early work on bootstrapping spatial data that properly takes account of geographical closeness through correlations that are high when the separation distance is small was due to Hall (1985). He proposed two different methods for bootstrapping spatial data. Both ideas start with the entire space D into k congruent subregions $D_1, D_2, \ldots, D_k$. The first method bootstraps the data and assigns bootstrap observation corresponding to index i to region D_j if the ith bootstrap observation is $\mathbf{X}_i^* = \mathbf{X}_j$. The second approach takes a bootstrap sample of the subgroups and then assigns all the data in that subgroup to the bootstrap sample to get the bootstrap estimate of the surface. The second method is very similar to the blocking idea of Kunsch (1989) for time series.

There are also semiparametric and parametric methods that are described in Cressie (1991, pp. 493–494 and p. 496, respectively). The semiparametric methods involve relating spatial data to explanatory variables. This allows the methods of Freedman and Peters (1984) and Solow (1985) to be applied to spatial data.

7.1.2 Asymptotics for Spatial Data

In order to study the properties of resampling methods for randomly generated spatial data, we must know the sampling mechanism that generates the original data because that is the mechanism that the resampling procedure needs to mimic. Different results can occur if we use different mechanisms. To study the asymptotic properties, we need a framework for the asymptotics in the spatial data setting. This setting will lead us to the extension of the block bootstrap to spatial data.

The main difference between spatial data and time series data is that in a time series, time is the natural order, and asymptotic theory is based on increasing time or the number of observations equally spaced in time. So for time series, the asymptotic direction refers to increasing time toward infinity or increasing the sample size n to infinity. But spatial data live in a two-dimensional or higher Euclidean space. So there is no obvious or unique direction to define the asymptotics.

So with spatial data, the concept "approaching infinity" must be defined to develop an asymptotic theory. But the definition for this concept is by no means unique, and the results depend on this choice. These concepts and examples are covered carefully in Cressie (1991, 1993). He described two basic paradigms for asymptotics. One is called increasing domain asymptotics and the other is called infill asymptotics. Cressie (1991, 1993) covered this is detail in chapter 5.

When the sites being sampled are separated by a fixed distance and the sample region grows without bound as the number of sites increase, we call it increasing domain asymptotics. This is the most common framework for spatial data. Using this form of asymptotics, we often get results analogous to the ones we get with time series. An example of an increasing domain would be a process observed over increasing nested rectangles on a rectangular grid of integer spacing.

On the other hand, infill asymptotics occurs when the sites are all located within a fixed boundary and the number of sites increases to infinity within the region. In this case, as the number of sites increase, the distance from one site to its nearest neighbor will decrease toward zero. This type of asymptotics is common with mining data and various other geostatistical applications.

There is a lot of literature on spatial data that show differences in inference between the infill and the increasing domain asymptotics. References dealing with this issue include Morris and Ebey (1984), Stein (1987, 1989), Cressie (1993), and Lahiri (1996). There is another form of spatial asymptotics that combines both of these approaches. One such example is called mixed increasing domain asymptotic structure. This is a case where the boundary of the region grows without bound but at the same time the number of sites is increasing fast enough to cause the distance between nearest neighbors to tend to zero.

7.1.3   Block Bootstrap on Regular Grids

To formally define spatial block bootstrap, we need a lot of machinery and terms that pertain strictly to spatial data. We will avoid this detail and instead focus on how the blocks are generated in the two dimensions. Lahiri (2003) provided many examples including a numerical illustration. When regular rectangular grids are used in two dimensions, Lahiri has shown consistency of the bootstrap. One such result is the consistency of the bootstrap estimate of variance (Lahiri, 2003, theorem 12.1, p. 295). This theorem requires several technical properties of stationary random fields, including a strong mixing condition.

7.1.4   Block Bootstrap on Irregular Grids

Lahiri also defined classes of spatial designs that lead to irregularly spaced grids. In theorem 12.6, Lahiri proved that the bootstrap estimates under irregularly space grids are consistent. This result is obtained under a spatial stochastic design for irregular grids that is covered in section 12.5.1. The approach is analogous to his results for long-range dependent series that we discuss briefly in Chapter 8.

7.2   SUBSET SELECTION IN REGRESSION

In classification or regression problems, we often have a large number of potential explanatory variables. Using all the variables is not likely to be the best solution because combinations of two or more of these variables can be highly correlated. In regression, this can cause ill-conditioning problems. Also, this approach is often considered to be overfitting because although including all variables leads to the best fit to the data, it usually is not the best model for prediction. So selecting a "best" subset is a way to do this.

There are many approaches because there are many different criteria that could be used to rank order the subsets. In linear regression, there are adjusted R-square, Mallow's Cp, Akaike information criterion (AIC), Bayesian information criterion (BIC), stepwise *F*-test, stochastic complexity, and more. Some of these approaches can be applied in nonlinear regression, time series, and other settings.

Here we present two examples. Gong was the first to apply the bootstrap as a tool to identify the most important variables to use in a model (a logistic regression for a medical application). We describe the approach and results in Section 7.2.1.

In Section 7.2.2, we present a very recent example where Gunter proposes the idea that when evaluating drugs in a clinical trial in addition to finding variables that influence the prediction of outcome, we should also include variables that help identify subsets where the chosen best treatments differ. The identification of such subsets may reveal that these variables are markers for prescribing different treatments depending on individual characteristics.

7.2.1 Gong's Logistic Regression Example

In Diaconis and Efron (1983), the idea of Gong was presented to the scientific community as an important new approach to variable selection. A detailed account of the approach and the example of Gregory's data can be found in Efron and Gong (1983) and Gong (1986).

In this example, Gong considered the data of 155 patients with acute or chronic hepatitis that had initially been studied by Dr. Gregory. Greogory was looking for a model that would predict whether or not a patient would survive the disease. In his sample, the outcome was that 122 survived the disease and 33 did not. Using his medical judgment, he eliminated six of the variables. He then applied forward logistic regression as a tool to eliminate 9 of the remaining 13 variables. Gregory was tempted to claim that the final set of four variables were the most important.

Gong questioned this conclusion, realizing that the stepwise procedure, which at least avoids the overfitting problem, could have by chance led to a very different set of variables that would predict equally well. The sensitivity of the choice of variables to slight changes in the data might show up by simulating a slightly perturbed data set. But this can also be accomplished by bootstrapping the entire stepwise regression procedure using a bootstrap sample as the perturbed data.

Gong generated 500 bootstrap replications of the data for the 155 patients. In some of the 500 cases, only one variable appeared in the final model. It could be ascites (the presence of fluid in the abdomen), the concentration of bilirubin in the liver, or the physician's prognosis. In other cases, as many as six variables appear in the final model. None of the variables emerged as very important by appearing in more than 60% of the bootstrap samples.

This was clearly a surprise to Dr. Gregory and showed what Gong expected (although the result may have been more dramatic than she had expected). So, one should be very cautious about attributing a level of importance to the variables that are included in a single stepwise logistic regression. This landmark work also showed how the bootstrap could be a very useful tool for determining which, if any, of the variables is truly important. One could carry this further to look at sensitivity of exploratory or data mining procedures on final results.

EXAMPLE 7.1 Stepwise Subset Selection

Consider the the artificial model

$$y = 0.5 + 0.5x + 0.2x^2 + 1.0z + 0.1xz + \varepsilon,$$

where $x = t$, $z = \sin(t)$, and $\varepsilon \sim N(0, 0.25)$ for t in the interval $[-5, +5]$. We wish to find this underlying model in the context of the more general form

$$y \sim x + z + x^2 + z^2 + xz + x^3 + z^3 + x^2z + xz^2,$$

using stepwise linear regression with the AIC. The following R script uses $B = 1000$ bootstrap resamples of 51 data to assess the frequency of occurrence of each of the nine possible terms in the fit using a method similar to that of Gong:

```
> ###EXAMPLE 7.2.1: Subset Selection
> set.seed(1) #for reproducibility
> t<- seq(-5,5,.2)
> n<- length(t) #number of data
> n
[1] 51
> x<- t
> z<- sin(t)
> ytrue<- 0.5 + 0.5*x + 0.2*x*x + 1*z + 0.1*x*z
> e<- rnorm(n, 0, 0.5)
> dat<- data.frame(x=x, x2=x*x, z=z, z2=z*z, xz=x*z, x3=x^3,
z3=z^3,
+ x2z=x*x*z, xz2=x*z*z, y=ytrue+e)
> head(dat)
      x     x2         z         z2         xz         x3
1  -5.0  25.00  0.9589243  0.9195358  -4.794621  -125.000
2  -4.8  23.04  0.9961646  0.9923439  -4.781590  -110.592
3  -4.6  21.16  0.9936910  0.9874218  -4.570979   -97.336
4  -4.4  19.36  0.9516021  0.9055465  -4.187049   -85.184
5  -4.2  17.64  0.8715758  0.7596443  -3.660618   -74.088
6  -4.0  16.00  0.7568025  0.5727500  -3.027210   -64.000
```

```
           z3        x2z        xz2         y
1  0.8817652  23.97311  -4.597679  3.166235
2  0.9885379  22.95163  -4.763251  3.317827
3  0.9811922  21.02650  -4.542140  2.550779
4  0.8617199  18.42302  -3.984405  3.502538
5  0.6620876  15.37460  -3.190506  2.598268
6  0.4334586  12.10884  -2.291000  1.743847
> round(cor(dat),4) #show correlations to 4 decimals
          x        x2         z        z2        xz        x3        z3
x    1.0000    0.0000   -0.2601    0.0000    0.0000    0.9168   -0.2848
x2   0.0000    1.0000    0.0000    0.2717   -0.9038    0.0000    0.0000
z   -0.2601    0.0000    1.0000    0.0000    0.0000   -0.5636    0.9544
z2   0.0000    0.2717    0.0000    1.0000   -0.2466    0.0000    0.0000
xz   0.0000   -0.9038    0.0000   -0.2466    1.0000    0.0000    0.0000
x3   0.9168    0.0000   -0.5636    0.0000    0.0000    1.0000   -0.5901
z3  -0.2848    0.0000    0.9544    0.0000    0.0000   -0.5901    1.0000
x2z -0.7116    0.0000    0.7488    0.0000    0.0000   -0.9255    0.7619
xz2  0.8661    0.0000   -0.3232    0.0000    0.0000    0.9161   -0.3663
y    0.6302    0.6765    0.1458    0.1952   -0.5878    0.4739    0.0986
          x2z        xz2         y
x     -0.7116     0.8661    0.6302
x2     0.0000     0.0000    0.6765
z      0.7488    -0.3232    0.1458
z2     0.0000     0.0000    0.1952
xz     0.0000     0.0000   -0.5878
x3    -0.9255     0.9161    0.4739
z3     0.7619    -0.3663    0.0986
x2z    1.0000    -0.8454   -0.2623
xz2   -0.8454     1.0000    0.5155
y     -0.2623     0.5155    1.0000
>
> #data fit
> require('MASS')
> fit1<- lm(y ~ x + z, data=dat) #starting model
> afit1<- stepAIC(fit1, scope=list(upper=~x+z+x2+z2+xz+x3+z3+
x2z+xz2, lower=~1),
+    trace=FALSE)
> summary(afit1)

Call:
lm(formula = y ~ x + z + x2 + xz + z3, data = dat)

Residuals:
```

```
     Min        IQ    Median        3Q       Max
 -1.12764  -0.25821  0.08315  0.20717  0.81946
```

Coefficients:

```
              Estimate  Std. Error  t value  Pr(>|t|)
(Intercept)    0.48959     0.13574    3.607  0.000774  ***
x              0.49090     0.02066   23.759   < 2e-16  ***
z              1.35342     0.26725    5.064  7.42e-06  ***
x2             0.20842     0.01758   11.859  1.92e-15  ***
xz             0.11686     0.06042    1.934  0.059399  .
z3            -0.55726     0.33028   -1.687  0.098480  .
---
```

```
Signif. codes: 0 '***' 0.001 '**' 0.01 '*' 0.05 '.' 0.1 ' ' 1
```

Residual standard error: 0.4161 on 45 degrees of freedom
Multiple R-squared: 0.9631, Adjusted R-squared: 0.959
F-statistic: 235 on 5 and 45 DF, p-value: < 2.2e-16

```
>
> #bootstrap to find common models
> nReal = 1000
> p<- rep(0,9) #proportions of each variable occurring
> for (iReal in 1:nReal) { #resample loop
+   ind<- sample(1:n, n, replace=TRUE) #bootstrap by indices
(cases)
+   bdat<- dat[ind,] #pick out rows
+   fit2<- lm(y ~ x + z, data=bdat) #starting model
+   afit2<- stepAIC(fit1, scope=list(upper=~x+z+x2+z2+xz+x3+z3+x2z
+xz2,
+     lower=~1), trace=FALSE)
+   s<- names(coef(afit2)) #get fit variables
+   m<- length(s)
+   for (j in 2:m) { #check terms for variables
+     if (s[j]=='x') {
+       p[1]<- p[1] + 1
+     } else if (s[j]=='x2') {
+       p [2] <- p [2] + 1
+     } else if (s[j]=='z') {
+       p [3] <- p [3] + 1
+     } else if (s[j]=='z2') {
+       p [4] <- p [4] + 1
+     } else if (s[j]=='xz') {
+       p [5] <- p [5] + 1
+     } else if (s[j]=='x3') {
+       p [6] <- p [6] + 1
```

```
+      } else if (s[j]=='z3') {
+        p [7] <- p [7] + 1
+      } else if (s[j]=='x2z') {
+        p [8] <- p [8] + 1
+      } else if (s[j]=='xz2') {
+        p [9] <- p [9] + 1
+      } else {
+        cat('Error! ', sprintf('%5d',m), sprintf('%5d',j), '\n')
+        cat(s)
+      }
+    }
+ }
> print('Variables: x, x*x, z, z*z, x*z, x^3, z^3, x*x*z, x*z*z')
[1] "Variables: x, x*x, z, z*z, x*z, x^3, z^3, x*x*z, x*z*z"
> print('True:    1  1   1   0   1   0   0   0      0')
[1] "True:    1 1   1   0   1   0   0   0      0"
> P<- p/nReal    #show proportions associated with variable terms
> P
[1] 1 1 1 0 1 0 1 0 0
>
```

Note that the final proportions indicate "all model terms were found (together with a spurious z^3 term) among the 1000 resample fits."

7.2.2 Gunter's Qualitative Interaction Example

Gunter et al. (2011) described a decision-making approach to variable selection. The key idea is that in the context of a randomized controlled clinical trial, the choice of treatment might differ on subset of the populations. Genetic markers are becoming popular as a tool in determining the best treatment for particular individuals.

Variable selection techniques in regression, logistic regression, or Cox proportional hazard models have traditionally concentrated on finding the variables that provide the best set of predictors of the outcome or response. Gunter et al. (2011) pointed out that some important variables can be missed this way. For example, we may be interested in variables such as a biomarker that have an interaction with the treatment regarding the outcome. This is best illustrated by figure 7.2.2 in their paper, which we reproduce as Figure 7.2.

At the top row of the figure, we see in (a) two parallel lines indicating no interaction at all. In (b), we see diverging lines. This is called a nonqualitative interaction. That is because although there is a larger difference between the two treatments at the high values of X_2 the decision of which treatment to use would not change because the lines never cross. In (c), we have a qualitative interaction because the lines cross, and so the decision on which drug to prescribe is opposite for high values of X_3 than what it is for low values of X_3.

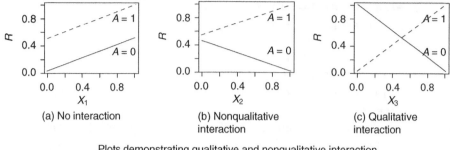

(a) No interaction   (b) Nonqualitative   (c) Qualitative
                               interaction   interaction

Plots demonstrating qualitative and nonqualitative interaction

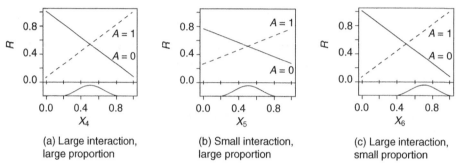

(a) Large interaction,   (b) Small interaction,   (c) Large interaction,
large proportion   large proportion   small proportion

Figure 7.2. Plots demonstrating usefulness factors of qualitative interactions.

An algorithm for selecting variables that includes both prediction and interaction variables is described in section 4.1 of the paper and is fairly complex. Involved in the process are some sophisticated selection criteria that uses the "Lasso" but involves the BIC to determine the penalty parameter. The authors illustrated their method through simulations and real data from the Nefazodone CBASP trial.

For this real data set, they augment the algorithm by bootstrapping the original data to measure the reliability of approach (or sensitivity to changes in the data). In this way, we see the ideas of Gong reappearing 24 years later. A method for setting thresholds for including an interaction variable is a second adjustment made to the algorithm for this practical problem. This algorithm augmented with the bootstrap takes 30 h to run using MATLAB and a 3-GHz Intel Xeon X5355 processor when there are 60 baseline covariates to consider along with 400 subjects in the database. This indicates that the method is extremely computer intensive.

7.3 DETERMINING THE NUMBER OF DISTRIBUTIONS IN A MIXTURE

Mixture distribution problems arise in various diverse ways. In the case of Gaussian mixtures, they have been used to identify outliers (see Aitkin and Tunnicliffe-Wilson, 1980) and to investigate the robustness of certain statistics as the underlying distribution departs from a Gaussian as in the case of Srivastava and Lee (1984) for the sample correlation coefficient when the distribution departs from a bivariate Gaussian distribution. It is being used increasingly in the field of cluster analysis as can be seen from the papers of Basford and McLachlan (1985a–d).

When Chernick worked in the aerospace industry on defense problems for the air force, targets and decoys both appeared to have feature distributions that could fit well to a mixture of multivariate Gaussian distributions. This could occur because the enemy could use different types of decoys or could apply materials to the targets to try to hide their signatures.

Discrimination algorithms based on mixture models could work once the mixture models are appropriately estimated from training data, but this assumes that the number of distributions in the mixture is known. The most difficult part of the problem is to determine the number of distributions in the mixture for the targets and the decoys. McLachlan and Basford (1988) applied a mixture likelihood approach to the clustering problem. One approach taken by McLachlan and Basford to deciding on the number of distributions in the mixture is to apply a likelihood ratio test employing bootstrap sampling.

The bootstrap is used because in most parametric mixture models, if we define λ to be the likelihood ratio statistic, $-2\log \lambda$ fails to have the usual asymptotic chi-square distribution because the required regularity conditions needed for the limit theorem fail to hold. Since the distribution of the test statistic under the null hypothesis is used, the idea is to approximate it with the bootstrap distribution for $-2\log \lambda$.

We now formulate the problem of likelihood estimation for a mixture model and then define the mixture likelihood approach. Then, we will look at a bootstrap test to determine the number of distributions in the mixture. We let $\mathbf{X}_1, \mathbf{X}_2, \ldots, \mathbf{X}_n$ be a sequence of p-dimensional random vectors. Each $\mathbf{X}_j$ is a random vector from a multivariate distribution G, which is a finite mixture of k probability distributions F_1, $F_2, \ldots, F_k$, which represent these distributions, and $\pi_1, \pi_2, \ldots, \pi_k$, which represent their proportions in the mixture. Consequently, we require

$$\sum_{i=1}^{k} \pi_i = 1, \pi_i \geq 0 \text{ for each } i = 1, 2, \ldots, k.$$

From this definition, we also see that the cumulative distribution G is related to the mixing distributions by $G(\mathbf{x}) = P[\mathbf{X}_i \leq \mathbf{x}] = \sum_{i=1}^{k} \pi_i F_i(\mathbf{x})$, where, by "$\leq$" we mean each component of the p-dimensional vector on the left-hand side is less than or equal to the corresponding component on the right-hand side. If we further assume that all the F_is are differentiable then G is also, and we can denote the relationship between the densities by $g_\varphi(x)$, the density of G parameterized by φ and $f_{i,\varphi}(x)$, the derivative of F_i also parameterized by φ, where φ is the vector $(\pi, \theta)^T$ and $\pi = (\pi_1, \pi_2, \ldots, \pi_k)^T$, and θ is a vector of unknown parameters that define the distributions F_i for $i = 1, 2, \ldots, k$.

If we assume that all the F_i's come from the same parametric family of distributions, there will be an identifiability problem because any rearrangement of the indices will not change the likelihood. This identifiability problem can be removed by specifying that $\pi_1 \geq \pi_2 \geq \pi_3 \geq \ldots \geq \pi_k$. With the ordering problem eliminated, maximum likelihood estimates of the parameters can be obtained by solving the system of equations:

$$\partial L\varphi/\partial\varphi = 0, \tag{7.1}$$

where L is the log of the likelihood function and φ is the vector parameter as previously defined.

The method of solution is the estimation and maximization (EM) algorithm of Dempster et al. (1977), which was applied earlier in specific mixture models by Hasselblad (1966, 1969), Wolfe (1967, 1970), and Day (1969). They recognized that through manipulation of Equation 7.1 above we define the a posteriori probabilities

$$\tau_{ij}(\varphi) = \tau_i(x_j : \varphi) = \text{probability that } x_j \text{ corresponds to } F_i$$

$$= \pi_i f_{i,\varphi}(x_i) \Bigg/ \left\{ \sum_j \pi_j f_{j,\varphi}(x_j) \right\} \text{ for } i = 1, 2, \ldots, k,$$

then

$$\hat{\pi}_i = \sum_{j=1}^{n} \hat{\tau}_{ij}/n \text{ for } i = 1, 2, \ldots, k$$

and

$$\sum_{i=1}^{k} \sum_{j=1}^{n} \hat{\tau}_{ij} \partial \log f_{j,\hat{\varphi}}(x_j)/\partial\hat{\varphi} = 0.$$

There are many issues related to a successful application of the EM algorithm for various parametric estimation problems including the choice of starting values. If the reader is interested in the technical details, he or she can consult McLachlan and Krishnan (1997) or McLachlan and Basford (1988).

One approach of McLachlan and Basford (1988) as mentioned previously applies the generalized likelihood ratio statistic under the null hypothesis. The likelihood ratio statistic λ is then used to calculate $-2\log \lambda$. Since the regularity conditions necessary to prove the asymptotic chi-square result fail in this case, a bootstrap distribution under the null hypothesis is used for the statistical test.

As a simple example, consider the test that assumes a single Gaussian distribution under the null hypothesis compared with a mixture of two Gaussian distributions under the alternative. In this case, the mixture model is

$$f_\varphi(x) = \pi_1 f_{1\varphi}(x) + \pi_2 f_{2\varphi}(x), \text{ where } \pi_2 = 1 - \pi_1.$$

Under the null hypothesis, $\pi_1 = 1$ and the solution is on the boundary of the parameter space. This shows why the usual regularity conditions fail in this case.

In certain special cases, the distribution of $-2\log \lambda$ under the null hypothesis can be derived when the solution under the null hypothesis falls on the boundary of the parameter space. For the reader interested in more detail, see McLachlan and Basford (1988, pp. 22–24) who provided specific examples.

In general, we will consider a nested model where, under the null hypothesis, there are k distributions in the mixture. Under the alternative hypothesis, there are $m > k$ distributions that include the original k distributions assumed under the null hypothesis. The bootstrap prescription takes the following steps:

1. Compute the maximum likelihood estimates under the null hypothesis.
2. Generate a bootstrap sample of size n, where n is the original sample size and the samples are assumed to come from the set of k distribution under the null hypothesis that was determined by maximum likelihood estimates in step 1.
3. Calculate $-2\log \lambda$ based on the value of λ determined in step 2 for the bootstrap sample.
4. Repeat steps (2) and (3) many times.

The bootstrap distribution of $-2\log \lambda$ is estimated by the Monte Carlo approximation, and since it is important for inference to have accurate estimates of the tails of the distribution, McLachlan and Basford recommended at least 350 bootstrap iterations. Given the approximate α level for the test, suppose in general we have m bootstrap iterations, and if $\alpha = 1 - j/(m + 1)$, then reject H_0 if $-2\log \lambda$ based on the original data exceeds the jth smallest value from the m bootstrap iterations.

For the case where we choose between a single Gaussian distribution and a mixture of two distributions, McLachlan (1987) used simulations to show how the power improves as m increases from 19 to 99. Several applications of this test applied to clustering problems can be found in McLachlan and Basford (1988).

7.4 CENSORED DATA

The bootstrap has also been applied to some problems involving censored data. Efron (1981) considered the determination of standard errors and confidence interval estimates for parameters of an unknown distribution when the data are subject to right censoring. He applied the bootstrap to the Channing House data first analyzed by Hyde (1980). Additional coverage of the approach is given in Efron and Tibshirani (1986).

Channing House is a retirement center in Palo Alto, CA. The data that have been analyzed consist of 97 men who lived in Channing House over the period from 1964 (when it first opened) to July 1, 1975 (when the data were collected and analyzed). During this time frame, 46 of the men had died, and the remaining 51 must be right censored on July 1, 1995.

The Kaplan–Meier curve (Kaplan and Meier, 1958) is the standard nonparametric method to estimate survival probabilities over time for right-censored data. The Greenwood formula for approximate confidence intervals is the most commonly used method to express the uncertainty in the Kaplan–Meier curve (although Peto's method is sometimes preferred by regulatory authorities such as the U.S. Food and Drug Administration [FDA]).

Efron (1981) compared a bootstrap estimate of the standard error for the Kaplan–Meier curve with the Greenwood estimate on the Channing House data. He found that the two methods provided remarkably close results. He also looked at bootstrap estimates for median survival and percentile method confidence intervals for it (not available by other nonparametric methods).

Right-censored survival data are usually coded as a bivariate vector $(x_i, d_i)^T$, where x_i is the observed event time for the ith patient and d_i is an indicator usually 1 for an outcome event and 0 for a censored event. In the case of the Channing House data, x_i is measured in months. So the pair (790, 1) would represent a man who died at age 790 months and (820, 0) would be a resident who was alive at age 820 months on July 1, 1975.

Efron and Tibshirani (1986) provided a statistical model for survival and censoring (see pages 64–65) that is used to generate the bootstrap data. They also looked at an approach that avoids the censoring mechanism and simply provides a bivariate distribution F for the pairs and simply samples with replacement from F. They found that either method provides essentially the same results. This insensitivity of the result to the censoring mechanism by the approach that uses F is reminiscent of the robustness of the use of bootstrapping vectors in the regression to the form of the regression model.

The work of Akritas (1986) is another key paper on bootstrapping censored data, specifically the Kaplan–Meier curve. This work is along the lines of Efron (1981) and Reid (1981).

7.5   *P*-VALUE ADJUSTMENT

In various studies where multiple testing is used or in clinical trials where interim analyses are conducted or multiple end points are used, adjustment for multiplicity can be important. Using the individual or "raw" *P*-value that ignores the fact that several different tests are conducted simultaneously can lead to very misleading results, indicating statistical significance in situations where there is none. The simultaneous inference approach that historically was first used for contrast testing in the analysis of variance can be applied to these situations. See Miller (1981) for thorough coverage up to that time. There are many recent developments covered in Hsu (1996). A simple bound on the *P*-value, the Bonferroni bound is universally applicable but consequently very conservative.

Westfall and Young (1993) provided resampling approaches including bootstrap and permutation methods for *P*-value adjustment. The early work is Westfall (1985), which considered multiple testing of binary outcomes. The approach is implemented in SAS using a procedure called PROC MULTTEST. It was first available in SAS Version 6.12. It has been updated and is available in the current SAS® Version 9.2.

7.5.1   The Westfall–Young Approach

In Westfall and Young (1993), they defined for each of the k tests that if we require the family-wise significance level (FWE) to be α, then for each $i = 1, 2, \ldots, k$, the adjusted *P*-value is defined to be

$$p_i^\alpha = \inf\{\alpha| H_i \text{ is rejected at FWE} = \alpha\}.$$

This means that it is the smallest significance level for which one still rejects H_i, the null hypothesis for the ith test given the specified multiple test procedure.

Next, they defined a single-step adjusted *P*-value. For the k tests, let $p_1, p_2, \ldots, p_k$ be the k raw *P*-values that have some unknown joint distribution. They then defined the one-step adjusted *P*-value p_{is}^α for test i to be the smallest *P*-value such that at least one of the k raw *P*-values $p_1, p_2, \ldots, p_k$ is no greater than p_i when the alternative hypothesis is true (i.e., all individual null hypotheses are false). In addition to this, Westfall and Young (1993) defined several other types of adjusted *P*-values, and for each, they provided resampling approaches to estimate it and particularly bootstrap estimates. For the single-step bootstrap adjusted *P*-value estimate, they defined the following algorithm:

Step 0: Set all counters to zero ($i = 1, 2, \ldots, k$).

Step 1: Generate bootstrap samples under the null hypothesis.

Step 2: Generate a *P*-value vector $\mathbf{p}^* = \left(p_1^*, p_2^*, p_3^*, \ldots, p_k^*\right)$.

Step 3: If $\min p_j^* \leq p_i$, where the minimum is taken over all $j = 1, 2, 3, \ldots, k$, then add 1 to the counter i for each i satisfying the inequality.

Step 4: Repeat steps 1 and 2 N times.

Step 5: Estimate p_{is}^α as $p_{is}^\alpha(N) = $ (value of counter i)/N.

7.5.2 Passive Plus Example

The Passive Plus steroid-eluting lead went through a clinical trial for approval using the same design as we presented for the Tendril DX lead. The issue we are discussing here is the use of interim analyses including the analysis of the primary end point (capture threshold at 3 months post implant). The FDA allows interim analyses for safety, and more recently, they can be used in a preplanned adaptive or group sequential design.

At the time of the submission, St. Jude chose to do interim analyses for a variety of reasons as discussed in Chernick (2007). These interim analyses were not preplanned to control the type I error so the FDA requires St. Jude to take a penalty for multiple looks. In this case, we were prepared to use Bonferroni, but if the Bonferroni bound turns out to be too conservative, we would present the bootstrap *P*-value adjustment as an alternative estimate of the appropriate penalty.

It turned out that the *P*-value for the final end point was less than 0.0001, and hence, even the Bonferroni adjustment would be no greater than 0.0004. So there was no need to apply the bootstrap adjustment. In the next example, we will see how the bootstrap can be useful when the Bonferroni bound is too conservative.

7.5.3 Consulting Example

In this example, a company conducted an international clinical trial for a medical treatment after seeing that they would not be able to get the enrollment in the United States as quickly as they had hoped. We label the United States as country E. Other countries in Europe were denoted A, B, C, and D. The treatment was the use of stents to clear blockage in the arteries. Restenosis was the primary end point. The goal was to show that the rate with the new stent was significantly lower than the historical control at 30%. The overall results are shown in Table 7.1.

We note that the sample sizes for countries A and C are a lot smaller than for the other countries, with country D having the largest number enrolled. It is interesting to note that the 22% in the United States appears to be better than for the historical control of 30%. There also seems to be vast differences in the rates across the countries.

Although the total sample size met their expectations and the rate in the United States was what they hoped for, the overall rate was above 30% and they could not get approval. In consulting with the client, his first thought was maybe the difference between countries could be statistically significant, and then if a good reason could be found why the rate is so much lower in the United States, a larger study in the United States might lead to an approval.

Ironically, had they stayed with the original plan and recruited all 550 patients in the United States with the 22% rate holding up, they probably would have gotten an approval. One could argue that the historical control was based on U.S. data so that it only makes sense to compare with U.S. patients if the U.S. surgeons, for example, generally use better surgical techniques or have better equipment than their European counterparts.

The first step in this would be to compare the U.S. results with each of the others, pairwise, to see if the United States has a statistically significantly lower restenosis rate. But this involves four applications of Fisher's exact test and an adjustment for multiplicity is needed. We applied the Bonferroni bound and compared it with the bootstrap method described in this chapter. Table 7.2 shows the comparison along with raw P-values.

PROC MULTTEST in SAS was used to calculate the bootstrap P-values. It is noticeable how much more conservative Bonferroni is relative to the bootstrap. Compared with B, the raw P-value is so low that we see the difference is highly significant for E versus B regardless of the method of comparison. For E versus C and D, the raw P-value is too high, indicating that if there truly is a 7% improvement in the United States, a much larger sample size would be required to detect it. In the case of E versus A, the Bonferroni bound gives a nonsignificant result even at the 0.10 level, but the bootstrap shows a significant difference at the 0.10 level. Although difference in

TABLE 7.1. Comparison of Treatment Failure Rates (Occurrences of Restenosis)

Country	Failure rate (restenosis rate)
	% (proportion)
A	40% (18/45)
B	41% (58/143)
C	29% (20/70)
D	29% (51/177)
E	22% (26/116)

TABLE 7.2. Comparison of P-Value Adjustments to Pairwise Fisher Exact Test Comparing Treatment Failure Rates (Occurrences of Restenosis)

Countries	Raw P-value	Bonferroni P-value	Bootstrap P-value
E versus A	0.0307	0.1229	0.0855
E versus B	0.0021	0.0085	0.0062
E versus C	0.3826	1.0	0.7654
E versus D	0.2776	1.0	0.6193

restenosis rate was 18%, nearly the same as for E against B, the higher P-value is due to there only being 45 patients recruited from country A compared with 143 for country B and 116 for country E. Increasing the sample size in A to 125 or more may have led to a significant test result similar to E versus B.

EXAMPLE 7.2 Consulting Example

The following R script estimates the adjusted P-values for the Fisher "exact" tests in the consulting example:

```
> ###EXAMPLE 7.5.3: Consulting Example (P-value adjustment)
> n<- c(45, 143, 70, 177, 116) #sample sizes
> ncum<- cumsum(n) #cumulative sum of sizes
> ncum
[1]  45 188 258 435 551
> x<- c(18, 58, 20, 51, 18) #failures
> N<- sum(n)      #total number data
> N
[1] 551
> X<- sum(x)      #total number failures
> X
[1] 165
>
> #Data vectors of 1's and 0's
> yA<- c(rep(1,18), rep(0,45-18))    #A data
> yB<- c(rep(1,58), rep(0,143-58))   #B data
> yC<- c(rep(1,20), rep(0,70-20))    #C data
> yD<- c(rep(1,51), rep(0,177-51))   #D data
> yE<- c(rep(1,26), rep(0,116-26))   #E data
>
> #Fisher raw P-values
> pRaw<- rep(0,4)
> pRaw[1]<- fisher.test(matrix(c(sum(yA==1),sum(yA==0),
+  sum(yE==1),sum(yE==0)),ncol=2))$p.value #A vs E
> pRaw[2]<- fisher.test(matrix(c(sum(yB==1),sum(yB==0),
+  sum(yE==1),sum(yE==0)),ncol=2))$p.value #B vs E
> pRaw[3]<- fisher.test(matrix(c(sum(yC==1),sum(yC==0),
```

```
+  sum(yE==1),sum(yE==0)),ncol=2))$p.value #C vs E
>  pRaw[4]<- fisher.test(matrix(c(sum(yD==1),sum(yD==0),
+  sum(yE==1),sum(yE==0)),ncol=2))$p.value #D vs E
>  pRaw #raw Fisher P-values
[1] 0.030713547 0.002135258 0.382611714 0.277606230
>  pBon<- 4*pRaw #Bonferroni-adjusted P-values
>  pBon
[1] 0.122854187 0.008541032 1.530446855 1.110424921

>  #boostrap P-value adjustment (Westfall-Young)
>  set.seed(1)      #for reproducibility
>  m<- rep(0,4)     #counters for P-value adjustment
>  pp<- rep(0,4)    #P-values for bootstrap resamples
>  y<- c(rep(1,X), rep(0,N-X)) #full set of data under null
>  nReal<- 10000
>  for (iReal in 1:nReal) { #resampling
+  yy<- sample(y, N, replace=TRUE) #resample null data
+  yyA<- yy[1:ncum[1]] #A resample under null
+  yyB<- yy[(ncum[1]+1):ncum[2]] #B resample under null
+  yyC<- yy[(ncum[2]+1):ncum[3]] #C resample under null
+  yyD<- yy[(ncum[3]+1):ncum[4]] #D resample under null
+  yyE<- yy[(ncum[4]+1):ncum[5]] #E resample under null
+  pp[1]<- fisher.test(matrix(c(sum(yyA==1),sum(yyA==0),
+     sum(yyE==1),sum(yyE==0)),ncol=2))$p.value #A vs E
+  pp[2]<- fisher.test(matrix(c(sum(yyB==1),sum(yyB==0),
+     sum(yyE==1),sum(yyE==0)),ncol=2))$p.value #B vs E
+  pp[3]<- fisher.test(matrix(c(sum(yyC==1),sum(yyC==0),
+     sum(yyE==1),sum(yyE==0)),ncol=2))$p.value #C vs E
+  pp[4]<- fisher.test(matrix(c(sum(yyD==1),sum(yyD==0),
+     sum(yyE==1),sum(yyE==0)),ncol=2))$p.value #D vs E
+  ppmin<- min(pp) #smallest
+  for (i in 1:4) { #bump counts
+     if (ppmin<=pRaw[i]) m[i]<- m[i]+1 #bump counter if larger
or equal
+     } #next i
+  }
>  pBoot<- m/nReal #Westfall-Young bootstrap-adjusted bootstrap
P-values
>  pBoot
[1] 0.0880 0.0066 0.7572 0.6140
```

Compare these Bonferroni-adjusted bootstrap *P*-values to those in Table 7.2.

7.6   BIOEQUIVALENCE

7.6.1   Individual Bioequivalence

Individual bioequivalence is a level of performance expected of a new formulation of
a drug to replace an existing approved formulation, or it could also be the performance

necessary to have a competitor's drug accepted as a generic for a marketed drug. In clinical trials, individual bioequivalence is difficult to access, and there are very few statistical techniques available to estimate it. Shao et al. (2000) developed a bootstrap procedure to test for individual bioequivalence. They also showed that their approach leads to a consistent bootstrap estimate.

Pigeot (2001) provided a nice summary of the jackknife and bootstrap, and pointed out the pros and cons of each. One example is given to demonstrate the virtues of the jackknife and the other for the bootstrap. She cautioned about the blind use of bootstrap approaches, pointing to a particular bootstrap approach that was proposed earlier in the literature, for individual bioequivalence that turned out to be inconsistent.

In the first example, a common odds ratio is estimated in a stratified contingency table analysis. She provided two jackknife estimators of the ratio that are remarkably good at reducing the bias of a ratio estimate. In the second example, she demonstrated how the bootstrap approach of Shao et al. (2000) is successfully applied to demonstrate individual bioequivalence.

For a period of time, the FDA was inclined to favor individual bioequivalence to average bioequivalence, and the bootstrap approach for individual bioequivalence was included in the Food and Drug Administration Guidance for Industry, Center for Drug Evaluation and Research (1997) for bioequivalence. We shall consider this example in detail. We should also point out that the FDA recently reversed itself and is now recommending that average bioequivalence can be used to establish bio-equivalence, but they are also very cautious about the treatment of outliers in crossover designs used in bioequivalence trials. The relevant new guidances are Food and Drug Administration Guidance for Industry, Center for Drug Evaluation and Research (2001, 2003).

In the example, two formulations of a drug are tested for individual bioequivalence, using bootstrap confidence intervals that are consistent. Efron's percentile method was used. Other higher-order bootstrap intervals could have been used in this application as Pigeot pointed out.

Demonstrating bioequivalence amounts to exhibiting equivalent bioavailability as determined by the log transformation of the area under the concentration versus time curve.

The difference between the measures for the old versus new formulation is used to evaluate whether or not the two formulations are equivalent. Calculating this measure for average bioequivalence is fairly straightforward and does not require bootstrapping. However, at that time, the FDA preferred individual bioequivalence, which requires comparing estimates of specific probabilities rather than simple averages.

The recommended design for this trial is called the "two-by-three crossover design." In this design, the old treatment is referred to as the "reference treatment" using the symbol R. The new formulation is considered the "treatment under study" and is denoted with the letter T. The design requires each patient to take the reference drug over two periods and the new treatment over one period. Randomization is required to determine the order of treatment. The possibilities are RRT, RTR, and TRR. It is called "two-by-three" because the subjects are only randomized to two of the three possible sequences, namely RTR and TRR. The sequence RRT is not used.

The following linear model is assumed for the pharmacokinetic response:

$$Y_{ijkl} = \mu + F_i + P_j + Q_k + W_{ljk} + S_{ikl} + \varepsilon_{ijkl},$$

where μ is the overall mean; P_j is the fixed effect for the jth period with the constraint $\Sigma P_j = 0$; Q_k is the fixed effect of the kth sequence with $\Sigma Q_k = 0$; F_l is the fixed effect for drug l (in this case, l is 1 or 2 representing T and R) $F_T + F_R = 0$; W_{ljk} is the fixed interaction between drug, period, and sequence; S_{ikl} is the random effect of the ith subject in the kth sequence under the lth treatment (drug); and the ε_{ijkl} is the independent random error. Under this model, individual bioequivalence is assessed by testing $H_0 : \Delta_{PB} \leq \Delta$ versus $H_1 : \Delta_{PB} > \Delta$, where $\Delta_{PB} = P_{TR} - P_{RR}$ and $P_{TR} = prob(|Y_T - Y_R| \leq r)$ and $P_{RR} = prob(|Y_R - Y_R'| \leq r)$, where $\Delta \leq 0$ and r are specified in advance, and Y_R' is the observed response on the second time the reference treatment is given. We would want P_{TR} to be high, indicating Y_T and Y_R are close to each other. We also expect Y_R and Y_R' to be close. So, ideally, $\Delta_{PB} = 0$. But for equivalence, it would be okay for Δ_{PB} to be slightly negative. How negative is determined by the specified Δ. Bioequivalence is achieved if the lower bound of the appropriate bootstrap confidence interval for Δ_{PB} is above the prespecified Δ.

Pigeot pointed out that the FDA guideline makes the general recommendation that the hypothesis test is best carried out by construction of an appropriate confidence interval. To claim bioequivalence, the lower limit of the confidence interval is greater than $-\Delta$ and the upper limit is less than Δ. In the context of equivalence testing, it is two one-sided simultaneous tests at level α that lead to an α-level test, but the corresponding confidence interval is a $100(1 - 2\alpha)\%$ level interval. So a 90% confidence interval is used to produce a 5% level test for bioequivalence. Pigeot pointed out that the recommended bootstrap test proposed by Schall and Luus (1993) is inconsistent.

She then showed how a minor modification makes it consistent. The proof of consistency comes from the theorem of Shao et al. (2000). The FDA guideline was changed in 2003 to require average bioequivalence instead of individual bioequivalence. See Patterson and Jones (2006) for details.

Without characterizing the type of bioequivalence being considered (apparently average bioequivalence because areas under the concentration of individual patients are summed), Jaki et al. (2010) considered ratios of composite areas under the concentration to establish bioequivalence of two treatments. The designs are not crossover, and so we assume that each patient only gets one of the two treatments. The authors did simulations to compare a variety of confidence interval estimates for the ratio of the areas under the concentration.

Among the estimates, a bootstrap-t interval was included. The best overall estimates were the bootstrap-t, Fieller's interval for ratios of independent normals, and a jackknifed version of the Fieller interval. In large samples, there was very little difference in the accuracy of coverage for the intervals. But bootstrap-t was the best in small sample sizes. This is probably because the composite areas under the concentration curve are not approximately normal when the sample size is small, but the bootstrap comes close to approximating the actual distribution for the ratio.

7.6.2 Population Bioequivalence

Czado and Munk (2001) presented simulation studies comparing various bootstrap confidence intervals to use to determine population bioequivalence. The authors pointed out that, aside from draft guidelines, the current guidelines in place for bioequivalence are for the use of average bioequivalence even though the statistical literature points toward the virtue of other measures namely, individual and population bioequivalence. These other measures appear to be more relevant to the selection of a generic drug to replace an approved drug. In an influential paper, Hauck and Anderson (1991) argued that prescribability of a generic drug requires consideration of population bioequivalence.

Essentially, population bioequivalence means that the probability distributions for the two treatments are practically the same. Munk and Czado (1999) developed a bootstrap approach for population equivalence. In Czado and Munk (2001), they considered extensive simulations demonstrating the small sample behavior of various bootstrap confidence intervals and, in particular, comparing the percentile method denoted PC against the BCa method.

The context is a 2×2 crossover design. The authors addressed the following five questions:

1. How do the PC and BCa intervals compare in small samples with regard to maintaining the significance level and power based on results in small samples?
2. What is the effect on the performance of PC and BCa intervals in both normal and non-normal population distributions?
3. What is the gain in power when a priori period effects are excluded?
4. Does the degree of dependence in the crossover trial influence the performance of the proposed bootstrap procedure?
5. How do these bootstrap tests perform compared with the test where the limiting variance is estimated?

Only bootstrap estimates that are available without a need to estimate the limiting variance were considered. This then ruled out the bootstrap percentile-t method. In general, the BCa intervals were superior to the PC intervals (which tended to undercover). There are gains in power when the effects can be excluded and when positive correlation within sequences can be assumed.

7.7 PROCESS CAPABILITY INDICES

In many manufacturing companies, process capability indices that measure how well the production process is doing measured against engineering specifications are estimated. These indices were used by the Japanese in the 1980s as part of the movement toward quality improvement that changed "made in Japan" from being a connotation for cheap workmanship to the standard of best in the world. At least this was the case for automobiles and electronic equipment. Much of the credit for the transformation in Japan should go to the teachings of Deming, Duran, and Taguchi.

In the 1990s, the U.S. manufacturers, particularly those in the automobile industry, turned to these same leaders to turn their business around after losing out to the Japanese automobile companies such as Toyota, Nissan, and Honda. These indices are now commonly used in most major U.S. industries. In medical device companies, these indices are now being incorporated in the product performance qualification process for these device companies.

Although it has been primarily a positive change, this movement has unfortunately also led to some misuses of the techniques. The buzz words of the day have been total quality management (TQM) and six sigma methods. Companies such as Motorola, General Electric (GE), and General Motors (GM) were highly successful at incorporating a six sigma methodology. However, there have also been many failures.

In companies that were successful, a highly skilled statistical research group existed. In other companies missing this infrastructure, the engineers with poor background in statistics learn the techniques and teach them to others. Unfortunately, the statistical theory for this is not taught in black belt classes. Some procedures that heavily depend on Gaussian distributional assumptions are blindly applied and consequently are misused and misinterpreted.

At Pacesetter (St. Jude Medical), a company that produces pacemakers and pacing leads, and Biosense Webster, a company that makes ablation catheters, Chernick experienced a number of cases where the device or component part for the company failed product qualification. This would occur even when there were no failures among the samples tested.

This problem always occurred because the number of tests was small and the data were not normally distributed. The threshold for acceptance based on either Gaussian tolerance intervals or a process capability index, C_{pk} (the ratio of the minimum of the distance from the lower specification limit to the mean divided by three times the standard deviation and the distance from the upper specification limit to the mean divided by three times the standard deviation), was set too high because it was based on the normality assumption that was not applicable. The distributions were either highly skewed or very short tailed. Nonparametric tolerance intervals would consequently be the appropriate approach. In the case of C_{pk}, its distribution under normality should be replaced by a nonparametric approach such as the bootstrap distribution for the C_{pk} estimate.

When process capability parameters are computed, numbers such as 1.0 and 1.33 or even 1.66 are taken as the standard. But this is because of an interpretation for Gaussian distributions. For other distributions, the Gaussian interpretation may be very wrong. In this case, the bootstrap can play a role by removing the requirement for normality. Kotz and Johnson (1993) were motivated to write their book on process capability indices to clear up confusion and provide alternative approaches. They pointed out that more than one simple index is needed to properly characterize a complex manufacturing process.

Non-Gaussian processes have been treated in a variety of ways. Many of these are discussed in Kotz and Johnson (1993, pp. 135–161). The seminal paper of Kane (1986) devotes only a short paragraph to this topic. Kane conceded that non-normality is a common occurrence in practice and that the confidence intervals based on normality assumptions could be sensitive to departure from that assumption. He stated that "Alas it is possible to estimate the percentage of parts outside the specification limits, either directly or with a fitted distribution. This percentage can be related to an equivalent capability for a normal distribution."

Kane's point is that, in some cases, a lower capability index for a particular distribution (like one with shorter tails than the normal distribution) has a lower percentage of its distribution outside the specification limits than the normal distribution with the same mean and standard deviation. Either the associated probability outside the limits should be used to judge the capability of the process or the index itself should be reinterpreted.

Kane's suggestion is to find the normal distribution with the same percentage outside the specification limits, to find its capability index, and to interpret the non-normal process to have that capability. This would allow normal distributions and non-normal distributions to be equitably compared. However, it arbitrarily uses the normal distribution for the standard and requires the others to adjust to that standard. It may be a practical solution though, since many quality managers think in terms of these standard index numbers rather than what they actually represent. So Kane's approach translates things in a fair way for these managers to understand.

Another approach is taken by Gunter in a series of four papers in the journal *Quality Progress*. In two of the papers (Gunter, 1989a, b), he emphasized the difference between "perfect" (precisely normal) and "occasionally erratic" processes (i.e., coming from a mixture of two normal distributions with mixing proportions p and $1 - p$). If p is close to 1, then the erratic part is due to the occasional sample from the second distribution, which is the more variable one and could have a different mean.

Gunter considered three types of distributions: (1) one highly skewed, a central chi-square with 4.5 degrees of freedom; (2) a heavy-tailed symmetric distribution, a central t with 8 degrees of freedom; and (3) a very short-tailed distribution, the uniform distribution on some interval $[a, b]$. Using these exact distributions, Gunter can determine the mean and standard deviation for the distributions and then evaluate the percentage of observations outside of plus or minus three standard deviations. The nonconformance is measured in part out of 1 million. Using shift and scale transformations, the distributions can be made to have a mean of 0 and a standard deviation of 1.

The results are trivial to obtain but strikingly different, indicating how important symmetry and the tail behavior are. In case (1), 14,000 out of 1 million would fall outside, but all would be above the $+3\sigma$ limit and none below the -3σ limit. For case (3), the uniform distribution, there are no cases outside the limits! This is because the tail of the distribution is so short that the probability density drops abruptly to 0 before reaching either $+3\sigma$ or -3σ.

For the t-distribution, we have 2000 above the $+3\sigma$ limit and 2000 below the -3σ limit. Contrast these numbers with the normal distribution, which has 1350 above the $+3\sigma$ limit and 1350 below the -3σ limit. So because of skewness and heavy tails in cases (1) and (2), respectively, the Gaussian outperforms these other two, but for shorter than Gaussian tails like the uniform, case (3) outperforms the Gaussian.

Gunter's approach shows that there can be a great difference in the percentage falling out of the limits, depending on the shape of the underlying distribution. In practical situations, the mean and variance will not be known, and so they must be estimated from the sample. This requires simulations. English and Taylor (1990) did this for normal, triangular, and uniform distributions. English and Taylor obtained results very similar to Gunter's. So estimating the parameters does not have much effect on the relative comparison.

We should now see why the companies ran into so much trouble in the use of their capability indices. They simply equated the proportion defective to the Gaussian capability index without realizing that they were making a restrictive assumption that their data probably did not satisfy. They set 1.33 as the goal for the index when the real goal should be low defective percentage. Instead of a direct estimate of this that could come from nonparametric tolerance intervals, the estimated C_{pk} is compared with the goal for Gaussian distributions often when the data have shorter than Gaussian tails.

Another aspect of this where the companies often go wrong is that they use a sample estimate for the index and treat it as though it is the true value of the parameter when in fact there is uncertainty associated with the estimate. The magnitude of this uncertainty depends on the sample size. Since engineering tests often take only 10–20 samples, particularly if the testing is destructive, the uncertainty is relatively large and cannot be ignored.

Kocherlakota et al. (1992) took a theoretical approach and derived the exact distribution of a slightly different capability index called Cp in two specific cases. Price and Price (1992) studied the behavior of the expected value of the estimated C_{pk} via simulation for a large variety of distributions. Other approaches could be to get confidence intervals for C_{pk} without making parametric assumptions. The bootstrap is made for this kind of work as we shall see after we go through some formal development. For a good historical account of the developments in process improvement and capability indices, see Ryan (1989, chapter 7). In general, capability indices are estimated from data and the distribution theory of these estimates is covered in Kotz and Johnson (1993).

The 1990s saw an explosion in the use and development of capability indices extending to the multivariate distribution (especially the bivariate case in practice) and including Bayesian approaches such as Bernardo and Irony (1996). A more detailed and pragmatic account of the various indices and their strengths and weaknesses is Kotz and Lovelace (1998). Although there are a number of improvements and more robust choices for capability indices, one of the most popular indices to use when both upper and lower specification limits are given is still the index we have been discussing, C_{pk}.

Let μ be the process mean and σ be the process standard deviation (we assume the process is "stable" and "under control"). Let LSL denote the lower specification limit and USL the upper specification limit. Then, C_{pk} is defined as the minimum of (USL $-\mu$)/(3σ) and $(\mu - \text{LSL})/(3\sigma)$ and is called a process capability index. In practice, μ and σ have to be estimated from the data. So let m and s denote the estimates of μ and σ, respectively. Then, for the estimate of C_{pk} let

$$\hat{C}_{pk} = \min\{(m-\text{LSL})/(3s), \text{USL}-m)/(3s)\}.$$

It is the distribution of this estimate that we need to obtain confidence intervals for C_{pk}. Bootstrap confidence intervals can be obtained from the bootstrap distribution for $\hat{C}_{pk}$.

In practice, the Monte Carlo approximation is used, and given the resulting bootstrap samples, any of the various bootstrap confidence interval methods can be used including the percentile method and the high-order methods such as BCa. From the bootstrap intervals, hypothesis tests can be constructed.

The various approaches are discussed in Kotz and Johnson (1993, pp. 161–164) and in the original work of Franklin and Wasserman (1991) and Price and Price (1992). In reviewing this work, Kotz and Johnson (1993) pointed to the work of Schenker to indicate that there are potential problems with the use of the bias corrected (BC) bootstrap. Although the authors apparently did not consider using BCa, it is certainly possible to do so. Also, although computationally intensive, a double bootstrap can be applied to any of the methods these authors did try.

It is now fairly easy to apply capability indices in the standard statistical packages. For example, SAS introduced a procedure called PROC CAPABILITY, which adds univariate analysis for the capability indices by modifying the UNIVARIATE procedure.

In the following example, we illustrate the use of bootstrap confidence intervals in capability studies based on the depth of a lesion in the heart from the use of a catheter ablation procedure. It is clear from the Shapiro–Wilk test, the stem-and-leaf diagram, the box plot, and the normal probability plots that the distribution of the data departs significantly from that of a normal distribution. This is enough to warrant the use of a robust or nonparametric estimate of the capability index. We illustrate how bootstrap confidence intervals could be applied to handle this problem.

Figure 7.3 is taken from figure 8.2 in Chernick (2007). It shows the SAS output from the UNIVARIATE procedure. These data came from data collected on beef hearts using a variety of catheters of the same type. These data were taken using catheter designated as number 12. The *P*-value for the Shapiro–Wilk test is 0.0066. The

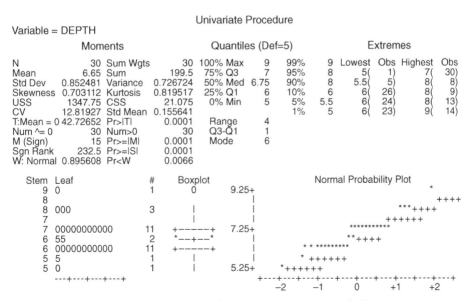

Figure 7.3. Summary statistics for Navistar DS lesion depth dimension.

stem-and-leaf diagram and the box plot show a positive skewness (sample estimate is 0.7031). For this catheter, a sample of 30 lesions was generated.

The purpose of these lesions is to destroy the nerve tissue that would otherwise stimulate the heart, causing the arrhythmia. If the lesion is effective, the electrical signal that causes the heart to contract is cut off, and ideally, only the normal pathway remains so that the patient will return to a normal sinus rhythm. If the lesion is not deep enough, the stimulus will remain or may recur in a short period of time. However, the lesion cannot be made too deep for then the heart could become perforated. This problem is hypothetical and is used to illustrate the bootstrap approach to capability index estimation, but it is very plausible that such an experiment would be done to make sure the depth of the lesion is within reasonable specification limits to ensure a successful ablation without perforation.

Tests of the catheter and the radio frequency (RF) generator are conducted on beef hearts so that the length, width, and depth of the lesion can be measured. This is often done to make sure that the devices are working properly but not necessarily to determine process capability as we are suggesting here.

For catheter 12, the average lesion depth was 6.65 mm with a standard deviation of 0.85 mm. The minimum lesion depth was 5.0 mm and the maximum was 9.0 mm. For illustrative purposes, we set the USL to 10 mm and the LSL to 4 mm. The target value is assumed to be 7.0 mm. Figure 7.4 taken from figure 8.3 in Chernick (2007) shows the output from the SAS PROC CAPABILITY procedure. It shows an estimate for C_{pk} and other capability indices. For the catheter 12 data, the C_{pk} estimate is 1.036.

We shall use the bootstrap to estimate the variance of this estimate and to construct approximate confidence intervals for C_{pk}. So we shall generate bootstrap samples of size 30 by sampling with replacement from the 30 lesion depths observed for catheter 12. Using the Resampling Stats, software I was able to construct bootstrap histograms for C_{pk} as well as Efron's percentile method for the confidence interval. Figure 7.5 exhibits the bootstrap histogram based on 10,000 bootstrap replications. The percentile

Capability Indices for Lesion Data

Variable = DEPTH   CAPABILITY

Moments				Quantiles (Def=5)			Extremes			
N	30	Sum Wgts	30	100% Max	9	99%	9	Lowest Obs	Highest Obs		
Mean	6.65	Sum	199.5	75% Q3	7	95%	8	5(1)	7(30)		
Std Dev	0.852481	Variance	0.726724	50% Med	6.75	90%	8	5.5(5)	8(8)		
Skewness	0.703112	Kurtosis	0.819517	25% Q1	6	10%	6	6(26)	8(9)		
USS	1347.75	CSS	21.075	0% Min	5	5%	5.5	6(24)	8(13)		
CV	12.81927	Std Mean	0.155641			1%	5	6(23)	9(14)		
T:Mean = 0	42.72652	Pr>|T|	0.0000	Range	4						
Sgn Rank	232.5	Pr>=|S|	0.0000	Q3-Q1	1						
Num ^= 0	30			Mode	6						
W: Normal	0.895608	Pr<W	0.0066								

Specifications				Indices			
LSL	4	% < LSL	0	CPL	1.036191	CPU	1.309902
Target	7			CP	1.173046	CPK	1.036191
USL	10	% > USL	0	K	0.116667	CPM	1.085148
		%Between	100	WARNING: Normality rejected			
				at α = 0.05			

Figure 7.4. Capability statistics for Navistar DS lesion depth dimension.

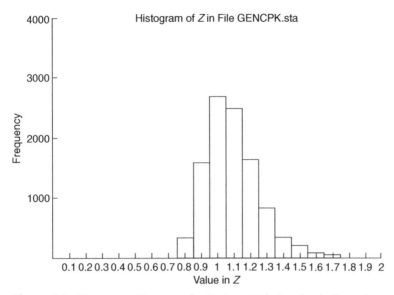

Figure 7.5. A bootstrap histogram for Navistar DS lesion depth dimension.

method confidence interval is [0.84084, 1.4608], showing a high degree of uncertainty in the estimate.

This may be a little surprising (particularly since there is only 0.5 resolution and many repeated values), but apparently, 30 samples is not enough to get a good estimate of the average lesion depth. Heavlin (1988) provided a good approximate two-sided confidence interval when the sample comes from a normal distribution. It is therefore interesting to look at the difference between Heavlin's two-sided 95% confidence interval and the bootstrap interval that we believe to be more reasonable due to the non-normality of the original data set. Applying Heavlin's method to the same lesion data as for the bootstrap gave [0.7084, 1.3649]. We observe that in comparison, the Heavlin interval is shifted to the left and is slightly wider. The Resampling Stats code that we used along with the output can be found in Chernick (2007, fig. 8.5, p. 165).

EXAMPLE 7.3 Heart Lesions
Using R,

```
> set.seed(1) #for reproducibility
> x<-
c(5.0,6.0,6.5,7.0,5.5,6.0,6.5,8.0,8.0,7.0,7.0,6.0,
8.0,9.0,6.0,6.0,6.0,

+ 6.0,7.0,7.0,6.0,7.0,6.0,6.0,7.0,6.0,7.0,7.0,7.0,
7.0) #lesion depths
> length(x)
[1] 30
```

```
> xm<- mean(x)
> xs<- sd(x)
> c(xm, xs, (10-xm)/(3*xs), (xm-4)/(3*xs))
#mean, std dev, CpkU, CpkL
  [1] 6.6500000 0.8524812 1.3099019 1.0361910
> qqnorm(x) #normal Q-Q plot
> qqline(x)
>
> require('boot')
> bootCpk<- function(x, i) {
+ m<- mean(x[i])
+ s<- sd(x[i])
+ CpkU<- (10-m)/(3*s)
+ CpkL<- (m-4)/(3*s)
+ return(min(CpkU, CpkL))
+ }
> bCpk<- boot(x, bootCpk, R=10000)
> boot.ci(bCpk)
BOOTSTRAP CONFIDENCE INTERVAL CALCULATIONS
Based on 10000 bootstrap replicates
CALL :
boot.ci(boot.out = bCpk)
Intervals :
Level Normal Basic
95% ( 0.682, 1.289 ) ( 0.620, 1.232 )
Level Percentile BCa
95% ( 0.840, 1.452 ) ( 0.769, 1.306 )
Calculations and Intervals on Original Scale
Warning message:
In boot.ci(bCpk) : bootstrap variances
needed for studentized intervals
```

Note that the BCa interval (0.769, 1.306) agrees more closely with Heavlin's (0.708, 1.365) than does the EP interval (0.840, 1.452).

7.8 MISSING DATA

Often in clinical studies, there will be some patients with incomplete or missing information. This can happen because the patient elects to drop out of the trial or in a quality-of-life survey certain questions are left blank. Most common statistical procedures require complete data. However, the EM algorithm allows likelihood inference to occur with some of the data missing.

Alternatively, there are a number of procedures available to impute values to all the missing data, and then the standard algorithms can be applied to the data that was imputed. However, when only one imputation is computed, the method underestimates the true variability. This is because the estimated variability obtained by the statistical procedure assumes all the data represent real observations. But now, there is a new component of variability that is due to the imputation. There are several ways to deal with this problem:

1. Use a mixed effects linear model to account for the variability due to imputing by accounting for it in the random effects.
2. Use the model with knowledge of the missing data mechanism to adjust the variance estimate.
3. Apply multiple imputation approach to simulate the variability due to imputation.

There are also approaches that fit a curve to the available data and use the curve to impute the missing values. When observations are missing at a follow-up between visits where data are available, we are using smoothing or interpolation. In the case of dropout, this approach is called extrapolation. There are many different ways to do this, and the success of the method will depend on the context. However, some crude methods such as "last observation carried forward" (LOCF) work only under very restrictive assumptions that are not usually met in practice.

Often, the data are not missing at random and so dropout could happen because of ineffective treatment, which may mean a negative trend in the measured value of the end point. But LOCF assumes that the next value would be expected to be the same as the previous one. Inspite of all of its drawbacks, LOCF was often accepted in the past before better imputation techniques were established. In some cases, it can be argued to be a conservative approach and as such may be acceptable to the FDA. But usually, it is very inaccurate, may not be conservative, and lacks an appropriate adjustment to variance estimates.

One of the early trials to demonstrate the inadequacy of LOCF, when compared with the multiple imputation approach, was a study at Eli Lilly that was presented by Stacy David at an FDA workshop in Crystal City Virginia in 1998. She compared multiple imputation methods with a variety of other techniques including LOCF using the Bayesian bootstrap (see Rubin, 1981 for the original work on the Bayesian bootstrap).

Also, at about the same time, AMGEN also presented advantages of using the multiple imputation approach in a real clinical trial. Since that time, many other studies have shown that LOCF is often an ineffective method and that there are better alternatives. A paper that motivated much of the research on multiple imputation methods in the pharmaceutical industry is Lavori et al. (1995). A clear modern treatment of the missing data problem is "Analysis of Incomplete Data," chapter 12, by Molenberghs et al. in Dmitrienko et al. (2007).

Recently, Todem et al. (2010) constructed a global sensitivity test for evaluating statistical hypotheses with models that have nonidentifiability problems. They had in particular, as a motivating example, the sensitivity analysis for missing data problems where the missingness mechanism depends on some unobservable outcomes. In section 3.2 of the paper, they used a real example, the fluvoxamine clinical trial, that was a longitudinal study with potentially nonrandom dropouts. Now, for global sensitivity analysis, the authors modeled the data and the missing mechanism jointly but cannot achieve a solution where the parameters for the data model and the missing mechanism model are jointly non-identifiable.

To solve the problem, the authors applied what they call an infimum test where the joint model includes random effects that are shared by the data and missingness components of the model. They constructed a bootstrap test on the infimum test statistic. The details of the bootstrap test results applied to the data from the fluvoxamine trial can be found in section 3.2 of the paper. This is an illustration of the potential of the bootstrap to help solve problems involving missing data where the missingness mechanism is nonignorable.

7.9 POINT PROCESSES

A point process is a collection of events over a particular time period. Examples include the time of failures for a medical device, the time to recurrence of an arrhythmia, the lifetime of a patient with a known disease, and the time of occurrence of an earthquake. This is just a small list of the possible examples.

Point processes can be studied in terms of the time between occurrences of events or the number of events that occur in a particular time interval. The simplest but not very commonly applicable example of a point process is the homogeneous Poisson process, which requires the following assumptions: (1) events occur rarely enough that no two events can occur over a small period of time and (2) the expected number of events in an interval of time T is λT where λ is the known rate or intensity for the occurrence of events. It is called a Poisson process because that number of events in a unit time interval has the Poisson distribution with parameter λ. A Poisson process also has the property that the time between events is independent of previous occurrences and has a negative exponential distribution with rate parameter λ.

The mean time between events for a Poisson process is λ^{-1}. The Poisson process is a special case of larger classes of processes including (1) stationary process, (2) renewal process, and (3) regenerative process. In addition to being able to define point processes over time, we can also define them in terms of other monotonically increasing parameters (such as cumulative costs). Point processes can also be defined in multidimensional space such as the epicenter of an earthquake in two dimensions. Such processes are called "spatial point processes."

There are also classes of nonstationary point processes. An example is an "inhomogeneous Poisson process." A detailed theoretical treatment can be found in Daley and Vere-Jones (1988) and Bremaud (1981). Murthy (1974) presented renewal processes along with medical and engineering examples. Bootstrap applications for

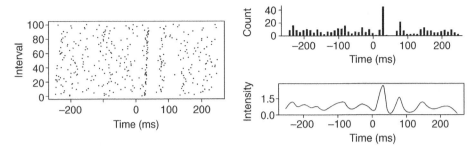

Figure 7.6. Neurophysiological point process. (From Davison and Hinkley (1997, figure 8.15, p. 418) with permission from Cambridge University Press).

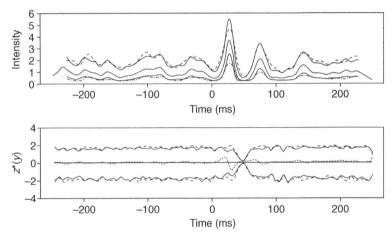

Figure 7.7. Confidence bands for the intensity of neurophysiological point process data. (From Davison and Hinkley (1997, figure 8.16, p. 421) with permission from Cambridge University Press.)

homogeneous and inhomogeneous Poisson processes are presented in Davison and Hinkley (1997, pp. 415–426). Davison and Hinkley defined some very simple approaches to bootstrapping point process, but they relied on restrictive assumptions.

An approach they applied to inhomogeneous Poisson processes is to smooth the estimate of the rate function based on the original data and then generate realizations of the process using this estimate of intensity. This is another example of a parametric bootstrap. Davison and Hinkley applied this approach to a point process of neurophysiological data. The whole process just described is illustrated in Figures 7.6 and 7.7 taken with permission from figures 8.1.5 and 8.1.6 of Davison and Hinkley (1997, pp. 418–421).

The data came from a study by Dr. S. J. Boniface from the Clinical Neurophysiology Unit and Radcliff Infirmary in Oxford, England. The data described a human response to a stimulus. The stimulus was applied 100 times and the points on the left portion of Figure 7.6 are represented on an intensity versus time curve over a 250-ms interval for which the response (firing of a motoneuron) was observed. Based on theoretical considerations, the process is obtained by superimposed occurrence and a rescaled kernel estimate of the intensity function $\lambda(t)$. Using the bootstrap, we are able to put approximate two-sided 95% confidence bands on the intensity at each given t. See Figure 7.7.

7.10   BOOTSTRAP TO DETECT OUTLIERS

For outlier detection, Mallows was the first to propose the use of influence functions. His work was documented in an unpublished paper but not was highlighted by Gnanadesikan (1977) in the context of multivariate outliers, using bivariate correlation as a prime example. In the late 1970s, Chernick was working on the validation of energy data. While outliers are not always errors, they are important as they can tell you unexpected things about your data. So an important part of validating data is to search for outliers.

In the context of energy data validation, a major goal was to evaluate the data in terms of their intended use and potential validity but not unanticipated uses. To make this concrete statistically, Chernick decided to formulate the intended use in terms of estimating statistical parameters from the data. Chernick (1982) derived the influence function for the multiple correlation coefficient, and illustrated how the bands of constant influence can be estimated at given points in the data space to indicate specific directions in two dimensions using these contours for the bivariate correlation. These contours turn out to be hyperbolae.

The influence function essentially measures how a single observation **x** in Euclidean space affects the estimate of the parameter of interest. So it can essentially tell you for any observation how much the estimate changes from the value of the parameter estimate obtained when it is included, compared with the value of the parameter estimate when it is left out. This is a very natural way to "look at" or "define" outliers in multivariate data as they relate to an important parameter. In the data validation context, you would identify outliers in terms of their effect on a list of parameters associated with the intended or anticipated potential use of the data.

The bootstrap is related to influence functions as pointed out by Efron (1979). So it is natural to think that the bootstrap could play a role in detecting outliers. Singh recognized this, and Singh and Xie (2003) introduced a graphic tool based on bootstrapping, which they called the "bootlier plot." It is based on the asymptotic result that a typical bootstrap sample (because of repeating observations) leaves out 36.8% of the data from the original sample. So, if outliers are present, many bootstrap samples will leave out the outliers or at least some of the outliers. Parameter estimates will change dramatically when the outliers are left out compared with when they are included. So the bootlier plot will show which bootstrap samples contain outliers, and those cases will identify the observations left in the bootstrap sample that are outliers.

In particular, Singh and Xie picked the mean minus the trimmed mean as the estimate to look at on the plot. The idea is that when the outliers are missing, the mean and trimmed mean will be nearly equal. But because the trimmed mean is robust and the mean is not, the estimate of the mean minus the trimmed mean will tend to be large when the outlier is included.

In the specific context of linear regression by least squares, Martin and Roberts (2006) proposed a bootstrap approach to estimate the null distribution of studentized residuals to test for outliers. It is well known that the studentized residuals are much better for graphically identifying outliers than the ordinary residuals. Martin and Roberts used the bootstrap distribution in order to avoid parametric assumptions.

Other than the fact that these two papers consider using the bootstrap to detect outliers, they have nothing in common. Singh and Xie exploited the sensitivity of bootstrap estimates to outliers and applied the bootlier graphs to point out the outlier. On the other hand, Martin and Roberts used the studentized residuals that are known to be useful for detecting outliers in least squares linear regression, a specialized solution. The bootstrap is used to determine critical values for the test.

7.11 LATTICE VARIABLES

Lattice variables are variables that are defined on a discrete set and so they apply to various discrete distributions including the binomial. In the papers, Singh (1981) and Bickel and Freedman (1981) required non-lattice variables to get consistency of the mean. This is illustrated in Lahiri (2006, p. 235) in theorem 11.1.

Other nice properties such as second-order consistency not only hold under the conditions of theorem 11.1 but also require the distribution to be non-lattice. For the lattice case, Lahiri (2006, p. 237) demonstrated the consistency property but the rate of convergence is no better than the simple normal approximation. So, although Lahiri was able to develop some theory for lattice variables, there does not seem to be a compelling reason to prefer the bootstrap over other methods, and there do not seem to be applications of the bootstrap on lattice variables in the literature.

7.12 COVARIATE ADJUSTMENT OF AREA UNDER THE CURVE ESTIMATES FOR RECEIVER OPERATING CHARACTERISTIC (ROC) CURVES

In recent years, it has become common for diagnostic tests to be evaluated by the behavior of their ROC curves. The "area under the ROC curve" (AUC) is commonly used as an index of the performance described by the ROC. An ideal ROC would have a waterfall shape staying as close to 1 as possible over a wide range of parameter values under the alternative hypothesis and dropping steeply to the significance level of the test as the parameters approach their value under the null hypothesis. In theory, the AUC for ROC could get arbitrarily close to 1 and could also be less than 0.5. But to be useful, an AUC must be greater than 0.5 since 0.5 can be achieved by random guessing as to which hypothesis is correct.

So a way to compare two diagnostic tests is to see which one produces the higher estimated AUC. In practice though, it is possible for the resulting curves to be

effected by covariates. This confounding by covariates could make the results misleading if the design used to estimate the AUC has different subjects taking one test than
those that take the other; a covariate imbalance could create bias that would make
the poorer diagnostic look like it is better than the other test. To handle this
problem, Yao et al. (2010) looked at parametric and nonparametric regression models
to describe the covariate effect. Once the models are fit to the data, the models
can be used for covariate adjustments that will make the comparison of AUCs fair. We
will focus on the nonparametric regression problem because the bootstrap is employed
there.

Yao et al. (2010) proposed the following regression models for two populations.
Individuals not having the disease are denoted by the random variable X and diseased
individuals Y. $\mathbf{Z}$ is a vector of covariate that could be confounding. The two nonparametric regression models are then

$$X \mid \mathbf{Z} = \mu_1(\mathbf{Z}) + \varepsilon_1\left(\sqrt{v_1(\mathbf{Z})}\right) \text{ and } Y \mid Z = \mu_2(\mathbf{Z}) + \varepsilon_2\left(\sqrt{v_2(\mathbf{Z})}\right),$$

where ε_1 and ε_2 are standardized error terms with mean 0 and variance 1 and independent of each other. Then, mean functions $\mu_1(\mathbf{Z})$ and $\mu_2(\mathbf{Z})$ are functions of the covariate
vector $\mathbf{Z}$ of an unspecified form. The variance functions $v_1(\mathbf{Z})$ and $v_2(\mathbf{Z})$ are also functions of the covariate vector $\mathbf{Z}$ of an unspecified form.

The idea of adjusting the AUC for covariates is not new. The authors are extending
work of Faraggi (2003) and Schisterman et al. (2004) who used normal regression
models to adjust the AUC index. But Yao et al. (2010) preferred the nonparametric
framework because the results from a parametric approach would be sensitive to model
misspecification in any of a variety of ways (incorrect functional form for the mean
and/or variance functions or incorrect form for the distribution of the error terms). A
fully nonparametric approach would not have this difficulty.

When the error terms in the model for $X|\mathbf{Z}$ and $Y|\mathbf{Z}$ are normally distributed, the
authors use the local polynomial modeling approach from the text by Fan and Gijbels
(1996) and are able to derive nice asymptotic properties including asymptotic normality
and strong consistency. The estimator that they used is called a covariate-adjusted
Mann–Whitney estimator (CAMWE) based on estimating the probability that $Y > X$
given $\mathbf{Z}$.

However, the real intent is to be fully nonparametric, and the assumption of
normal error distributions makes it semiparametric and still somewhat sensitive to
these assumptions. But for the fully nonparametric approach, the need for point-
wise confidence bands requires a bootstrap approach. In this case, a two-sided
symmetric 95% confidence interval is constructed using Efron's simple percentile
method. The authors admitted to not having theoretical support for the validity of
the bootstrap, but they did a variety of simulations using various complicated
regression functions to be fit using CAMWE along with sometimes very non-
normal residual distributions (e.g., lognormal for skewness and t for heavy tails).

The authors believed that the extensiveness of their simulations shows that the bootstrap can safely be applied.

In our presentation, we assumed **Z** to be a vector of covariates, whereas the authors derived their results only in the case of a single covariate but the extension to many covariates is practical and clear. However, each new covariate adds tremendously to the computer intensity of the method.

7.13   BOOTSTRAPPING IN SAS

Historically, there were not many ways to do bootstrapping in SAS, and researchers would do their own programming in S or more recently R. Insightful Corporation had the bootstrap as part of its SPlus software, which was produced by Tim Hesterberg. Resampling Stats, Inc., had its own software that did the basic bootstrap but only with Efron's percentile method confidence interval. In recent years, that software in its Excel version added the BCa confidence intervals.

SAS developed a program with the help of Westfall and Young to do bootstrap *P*-value adjustment called Proc MULTTEST, which had a bootstrapping algorithm imbedded in it. In SAS Version 9.2, Proc Survey Select (PSS) has a bootstrap sampling routine. Although PSS was designed to be efficient, the newly developed One-Pass, Duplicates-Yes (OPDY) by J. D. Opdyke was found to be generally faster than any of the other six algorithms that he compared it with. Paper by Opdyke published online for InterStat in October 2010 compares the speed of seven SAS algorithms to generate bootstrap samples:

1. OPDY;
2. PSS;
3. Hash Table, Proc Summary (HTPS);
4. Hash Table, Hash Iterator (HTHI);
5. Direct Access (DA);
6. Output–Sort–Merge (Out-SM);
7. Algorithm 4.8 (A4.8).

OPDY far exceeds PSS and the others with regard to real time speed. See Table 7.3 taken with permission from Opdyke (2010). Of course, this can be a little misleading because it is only a relative speed and if the actual speeds are very fast, a factor of 10 improvement may not be all that important. Also, since it is only the real time and CPU time for generating the bootstrap samples, this does not measure overall speed on a particular problem. If the problem is subset selection in regression, a linear mixed model, a time series model, or a generalized linear model, the bootstrapping has a SAS procedure imbedded in it, and these procedures can take up a lot of time. This could reduce or eliminate the advantage of OPDY.

TABLE 7.3. Comparisons from Table 1 of Opdyke (2010) Run Times and CPU Times Relative to OPDY (Algorithm Time/OPDY Time)

N^* per stratum	Number of strata	n^*	m^*	Real						CPU					
				PSS	A4.8	HTPS	HTIT	DA	Out-SM	PSS	A4.8	HTPS	HTIT	DA	Out-SM
10,000	2	500	500	5.3	10.2	9.0	4.5	8.8	7.5	2.6	10.1	3.3	2.1	3.8	6.0
100,000	2	500	500	23.3	83.3	6.2	4.6	7.3	9.7	18.0	81.4	7.3	2.4	3.1	4.4
1,000,000	2	500	500	31.1	121.2	2.5	1.9	1.4	4.2	25.0	119.3	1.2	1.2	0.7	1.8
10,000,000	2	500	500	42.2	164.9	2.5	2.7	0.7	7.6	33.4	162.0	1.4	1.4	0.4	1.8
10,000	6	500	500	8.4	16.8	10.0	6.8	13.8	14.3	5.8	16.6	4.5	3.6	6.0	7.6
100,000	6	500	500	23.7	84.0	4.7	4.3	7.4	8.7	18.9	82.7	2.8	2.5	3.3	4.4
1,000,000	6	500	500	37.3	145.0	2.3	2.1	4.4	4.5	30.4	143.2	1.4	1.4	0.8	2.2
10,000,000	6	500	500	39.8	370.8	‡	‡	7.8	8.8	32.5	349.1	‡	‡	0.4	1.8
10,000	12	500	500	11.4	23.7	17.9	28.0	18.1	16.8	8.5	23.1	5.8	4.6	8.0	9.6
100,000	12	500	500	31.4	96.3	9.4	9.0	8.7	8.6	24.7	94.2	3.2	2.7	3.4	5.0
1,000,000	12	500	500	47.0	160.7	3.8	3.3	10.1	4.2	38.1	155.8	1.7	1.6	0.9	2.2
10,000,000	12	500	500	45.0	‡	‡	‡	11.0	7.1	36.8	‡	‡	‡	0.4	1.8
10,000	2	1000	1000	5.3	7.3	7.2	5.7	11.2	12.4	3.6	7.0	3.7	2.8	5.3	7.2
100,000	2	1000	1000	25.7	80.4	8.5	7.4	14.3	16.0	19.9	78.1	4.7	3.6	6.4	8.4
1,000,000	2	1000	1000	50.0	183.9	3.3	2.5	8.9	4.4	39.9	179.7	1.9	1.6	1.6	2.7
10,000,000	2	1000	1000	65.2	1428.7	2.5	2.3	9.1	6.2	52.1	1218.7	1.2	1.1	0.5	1.6
10,000	6	1000	1000	8.2	10.7	16.3	17.9	16.9	26.8	5.3	10.4	5.4	4.3	7.6	10.7

N		n	m												
1,000,000	6	1000	1000	56.8	214.4	5.2	4.8	4.2	8.5	46.0	211.3	2.2	2.0	1.7	3.2
10,000,000	6	1000	1000	63.6	†	‡	‡	9.5	6.9	51.0	†	‡	‡	0.5	1.5
10,000	12	1000	1000	11.2	15.8	19.6	18.0	24.3	37.8	8.0	15.3	8.4	6.5	10.7	16.4
100,000	12	1000	1000	15.1	47.7	6.1	6.0	8.7	12.2	11.8	47.2	2.9	2.2	3.6	5.3
1,000,000	12	1000	1000	61.1	213.6	4.4	4.8	10.1	8.8	50.4	210.6	2.3	2.0	1.8	3.3
10,000,000	12	1000	1000	63.1	†	‡	‡	10.6	6.0	51.7	†	‡	‡	0.7	1.8
10,000	2	2000	2000	10.6	8.8	27.0	25.6	28.6	49.2	7.6	8.6	9.3	7.1	12.4	22.2
100,000	2	2000	2000	14.2	39.1	12.5	14.8	14.4	20.8	10.8	38.2	4.5	3.5	5.8	10.5
1,000,000	2	2000	2000	45.6	167.9	5.3	5.6	6.5	11.6	36.5	166.2	2.6	2.1	2.6	4.8
10,000,000	2	2000	2000	87.2	†	3.4	2.8	8.5	6.1	69.4	†	1.3	1.1	0.7	1.8
10,000	6	2000	2000	13.5	10.0	29.6	20.8	32.0	92.9	8.5	9.8	11.0	8.1	13.4	26.1
100,000	6	2000	2000	23.9	63.7	19.5	17.0	23.1	50.5	18.0	62.8	7.4	5.9	8.9	17.6
1,000,000	6	2000	2000	62.6	230.3	7.4	6.6	8.8	32.1	50.0	224.2	3.6	3.0	3.3	7.0
10,000,000	6	2000	2000	85.9	†	‡	‡	7.8	12.7	68.2	†	‡	‡	0.7	1.8
10,000	12	2000	2000	13.1	10.2	27.1	24.3	33.1	85.6	9.1	10.0	11.6	8.6	14.0	28.1
100,000	12	2000	2000	26.4	63.5	17.3	15.9	25.4	55.4	18.5	62.5	8.2	6.2	10.1	19.0
1,000,000	12	2000	2000	62.2	207.9	7.0	6.6	8.1	21.0	49.7	205.2	3.6	2.9	3.0	6.2
10,000,000	12	2000	2000	50.2	†	‡	‡	6.0	3.8	40.5	†	‡	‡	0.4	1.0

*N is the number of Monte Carlo replications in each stratum, n is the size of the original sample, and m is the size of the bootstrap sample.
†Excessive time.
‡Program crashed.

7.14   HISTORICAL NOTES

The application of the bootstrap to kriging that we presented is due to Switzer and Eynon, first described in the literature in Diaconis and Efron (1983), referring to an unpublished Stanford document. Cressie (1991, 1993) provided bootstrap approaches to spatial data and continues to be the best source for information on resampling with regard to spatial data. The text by Lahiri (2003) covers many of the more recent development in spatial data analysis with respect to resampling methods. Hall (1985, 1988b) provided very thought-provoking accounts regarding the application of bootstrap approaches to spatial data.

There are many excellent texts that cover statistical analysis of spatial data, including Bartlett (1975), Diggle (1983), Ripley (1981, 1988), Cliff and Ord (1973, 1981), and Upton and Fingleton (1985, 1989). Matheron (1975) provided the important fundamental probability theory results for spatial data. The work of Mardia et al. (1979) is a general multivariate analysis text that includes treatment of spatial data.

The idea of applying the bootstrap to determine the uncertainty associated with statistical models that are obtained through variable selection procedures was first considered by Gong (1982). It is somewhat disappointing that this line of research has not been pursued much since 1986. Gunter et al. (2011) looked at new ideas for variable selection based on the ability of the variables to demonstrate interaction between two treatments and the variable. This affects decisions on prescribing the treatment. Bootstrap methods for evaluating the selection are used. Gong's work appears in Efron and Gong (1981, 1983), Gong (1982, 1986), and Diaconis and Efron (1983). Miller (1990) dealt with model selection procedures in regression. A general approach to model selection based on information theory is called stochastic complexity and is covered in detail in Rissanen (1989).

Applications of the bootstrap to determine the number of distributions in a mixture model can be found in McLachlan and Basford (1988). A simulation study of the power of the bootstrap test for a mixture of two normals versus a single normal is given by McLachlan (1987).

Work on censored data applications of the bootstrap began with Turnbull and Mitchell (1978) who considered complicated censoring mechanisms. The work of Efron (1981) is a seminal paper on this subject.

The high-order accurate bootstrap confidence intervals are provided in Efron (1987) and Hall (1988a), and they usually provide improvements over the earlier procedures discussed in Efron (1981). Reid (1981) provided an approach to estimating the median of a Kaplan–Meier survival curve based on influence functions.

Akritas (1986) compared variance estimates for median survival using the approach of Efron (1981) versus the approach of Reid (1981). Altman and Andersen (1989), Chen and George (1985), and Sauerbrei and Schumacher (1992) applied case resampling methods to survival data models such as the Cox proportional hazards model. There is no theory yet to support this approach.

Applications of survival analysis methods and reliability studies can be found in Miller (1981) and Meeker and Escobar (1998), to name two out of many. Csorgo et al. (1986) applied results for empirical processes to reliability problems.

The application of the bootstrap to P-value adjustment for multiplicity was first proposed by Westfall (1985) and was generalized in Westfall and Young (1993). Their approach used both bootstrap and permutation methods, which they got incorporated into the SAS software procedure MULTTEST. PROC MULTTEST has been available in SAS since Version 6.12. There is also some coverage of multiple testing resampling approaches in Noreen (1989).

Regarding bioequivalence/bioavailability, the relevant regulatory documents are Food and Drug Administration Guidance for Industry, Center for Drug Evaluation and Research (1997, 2001, 2003) and the European Agency for the Evaluation of Medical Products (2001, 2006). Recent textbooks addressing bioequivalence are Patterson and Jones (2006), Hauschke et al. (2007), and Chow and Liu (2009). The text by Chow and Liu (2009) goes into great detail about the history of bioequivalence studies and is now in a mature third edition. Their extensive bibliography includes a long list of regulatory documents directly or indirectly affecting bioequivalence studies. The issue of outliers in bioequivalence studies is covered in Chow and Liu (2009), and one of the most recent references on this subject is Schall et al. (2010).

Initial work on estimates of process capability indices is due to Kane (1986). The bootstrap work is described in Franklin and Wasserman (1991, 1992, 1994), Wasserman et al. (1991), and Price and Price (1992). A review of this work and a general and comprehensive account of process capability indices can be found in Kotz and Johnson (1993).

Point process theory is found in Bremaud (1981), Murthy (1974), and Daley and Vere-Jones (1988, 2003) and in the context of level crossings and their relationship to extremes in Leadbetter et al. (1983) and Resnick (1987).

Rubin introduced the Bayesian bootstrap in Rubin (1981) in the context of missing data. Dempster et al. (1977) became the seminal paper that introduced the EM algorithm as an important tool when dealing with incomplete data. A detailed recent account of the method is given in McLachlan and Krishnan (1997). The multiple imputation approach was developed by Rubin and is extensively covered in Rubin (1987) and discussed along with other advanced method in the text (Little and Rubin, 1987).

7.15 EXERCISES

1. In Example 7.1, the "true" model included the terms "x," "z," "x^2," and "xz," yet the stepAIC() function only found models including the terms "x," "z," "x^3," and "z^3."

   (a) What does this say about the ability of stepAIC() to find the real model underlying a set of data?

   (b) All 1000 resamples led to essentially the same model form. What features of the data could explain this?

   (c) What do you think the effect of sample size would be on the results of this simulation? For example, would you expect any different behavior if the number of data in the sample were doubled or tripled?

2. Repeat the simulation in Example 7.1, but using the BIC. (Hint: Use the parameter "$k = \log(n)$" in stepAIC().)

   (a) How do the results compare with those found using the AIC?

   (b) What does this imply to you?

3. Surimi data: Example 3.1: Using the Surimi data of Example 3.1 and the specification limits of 20–60 for the force, find the following:

   (a) the mean and standard deviation for the data;

   (b) the C_{pk} value for the data;

   (c) 95% confidence limits using the EP and BCa methods for C_{pk} using 10,000 resamples;

   (d) comment on the widths of the intervals estimated.

   Hint: Use the R script

```
set.seed(1) #for reproducibility
x<- c(41.28, 45.16, 34.75, 40.76, 43.61,
39.05, 41.20, 41.02, 41.33, 40.61, 40.49,
  41.77, 42.07, 44.83, 29.12, 45.59, 41.95,
45.78, 42.89, 40.42, 49.31, 44.01,
  34.87, 38.60, 39.63, 38.52, 38.52, 43.95,
49.08, 50.52, 43.85, 40.64, 45.86,
  41.25, 50.35, 45.18, 39.67, 43.89, 43.89,
42.16)
length(x)
xm<- mean(x)
xs<- sd(x)
USL<- 60
LSL<- 20
c(xm, xs, (USL-xm)/(3*xs), (xm-LSL)/(3*xs))
#mean, std dev, CpkU, CpkL
qqnorm(x) #normal Q-Q plot
qqline(x)
require('boot')
bootCpk<- function(x, i) {
m<- mean(x[i])
s<- sd(x[i])
CpkU<- (USL-m)/(3*s)
CpkL<- (m-LSL)/(3*s)
return(min(CpkU, CpkL))
}
bCpk<- boot(x, bootCpk, R=10000)
boot.ci(bCpk)
```

REFERENCES

Aitkin, M., and Tunnicliffe-Wilson, G. (1980). Mixture models, outliers, and the EM algorithm. Technometrics 22, 325–331.

Akritas, M. G. (1986). Bootstrapping the Kaplan-Meier estimator. J. Am. Stat. Assoc. 81, 1032–1038.

Altman, D. G., and Andersen, P. K. (1989). Bootstrap investigation of the stability of a Cox regression model. Stat. Med. 8, 771–783.

Bartlett, M. S. (1975). The Statistical Analysis of Spatial Pattern. Chapman & Hall, London.

Basford, K., and McLachlan, G. J. (1985a). Estimation of allocation rates in a cluster analysis context. J. Am. Stat. Assoc. 80, 286–293.

Basford, K., and McLachlan, G. J. (1985b). Cluster analysis in randomized complete block design. Commun. Stat. Theory Methods 14, 451–463.

Basford, K., and McLachlan, G. J. (1985c). The mixture method of clustering applied to three-way data. Classification 2, 109–125.

Basford, K., and McLachlan, G. J. (1985d). Likelihood estimation with normal mixture models. Appl. Stat. 34, 282–289.

Bernardo, J. M., and Irony, I. Z. (1996). A general multivariate Bayesian process capability index. Statistician 45, 487–502.

Bickel, P. J., and Freedman, D. A. (1981). Some asymptotic theory for the bootstrap. Ann. Stat. 9, 1196–1217.

Bremaud, P. (1981). Point Processes and Queues: Martingale Dynamics. Springer-Verlag, New York.

Chen, C., and George, S. I. (1985). The bootstrap and identification of prognostic factors via Cox's proportional hazards regression model. Stat. Med. 4, 39–46.

Chernick, M. R. (1982). The influence function and its application to data validation. Am. J. Math. Manag. Sci. 2, 263–288.

Chernick, M. R. (2007). Bootstrap Methods: A Guide for Practitioners and Researchers, 2nd Ed. Wiley, Hoboken, NJ.

Chow, S.-C., and Liu, J.-P. (2009). Design and Analysis of Bioavailability and Bioequivalence Studies, 3rd Ed. Chapman & Hall/CRC, Boca Raton, FL.

Cliff, A. D., and Ord, J. K. (1973). Spatial Autocorrelation. Pion, London.

Cliff, A. D., and Ord, J. K. (1981). Spatial Processes: Models and Applications. Pion, London.

Cressie, N. (1991). Statistics for Spatial Data. Wiley, New York.

Cressie, N. (1993). Statistics for Spatial Data. Revised Edition Wiley, New York.

Csorgo, M., Csorgo, S., and Horvath, L. (1986). An Asymptotic Theory for Emiprical Reliability and Concentration Processes. Springer-Verlag, New York.

Czado, C., and Munk, A. (2001). Bootstrap methods for the nonparametric assessment of population bioequivalence and similarity of distributions. J. Stat. Comput. Simul. 68, 243–280.

Daley, D. J., and Vere-Jones, D. (1988). An Introduction to the Theory of Point Processes. Springer-Verlag, New York.

Daley, D. J., and Vere-Jones, D. (2003). An Introduction to the Theory of Point Processes. Volume I: Elementary Theory and Methods, 2nd Ed. Springer-Verlag, New York.

Davison, A. C., and Hinkley, D. V. (1997). Bootstrap Methods and Their Applications. Cambridge University Press, Cambridge.

Day, N. E. (1969). Estimating the components of a mixture of two normal distributions. Biometrika 56, 463–470.

Dempster, A. P., Laird, N. M., and Rubin, D. B. (1977). Maximum likelihood from incomplete data via the EM algorithm (with discussion). J. R. Stat. Soc. B 39, 1–38.

Diaconis, P., and Efron, B. (1983). Computer-intensive methods in statistics. Sci. Am. 248, 116–130.

Diggle, P. J. (1983). The Statistical Analysis of Spatial Point Patterns. Academic Press, New York.

Dmitrienko, A., Chung-Stein, C., and D'Agostino, R. (eds.). (2007). Pharmaceutical Statistics Using SAS: A Practical Guide. SAS Institute, Inc, Cary, NC.

Efron, B. (1979). Bootstrap methods: Another look at the jackknife. Ann. Stat. 7, 1–26.

Efron, B. (1981). Censored data and the bootstrap. J. Am. Stat. Assoc. 76, 312–319.

Efron, B. (1987). Better bootstrap confidence intervals (with discussion). J. Am. Stat. Assoc. 82, 171–200.

Efron, B., and Gong, G. (1981). Statistical thinking and the computer. In Proceedings Computer Science and Statistics, Vol. 13 (W. F. Eddy, ed.), pp. 3–7. Springer-Verlag, New York.

Efron, B., and Gong, G. (1983). A leisurely look at the bootstrap, the jackknife and cross-validation. Am. Stat. 37, 36–68.

Efron, B., and Tibshirani, R. (1986). Bootstrap methods for standard errors, confidence intervals and other measures of statistical accuracy. Stat. Sci. 1, 54–77.

English, J. R., and Taylor, G. D. (1990). Process capability analysis—A robustness study. Master of Science. University of Arkansas, Fayetteville, NC.

European Agency for the Evaluation of Medical Products. (2001). The investigation of bioavailability and bioequivalence. Note for guidance. Committee for Proprietary Medicinal Products (CPMP), London.

European Agency for the Evaluation of Medical Products. (2006). Questions and Answers on the bioavailability and bioequivalence guideline. EMEA/CHMP/EWP/40326, London.

Fan, J., and Gijbels, I. (1996). Local Polynomial Modelling and Its Applications. Chapman & Hall, London.

Faraggi, D. (2003). Adjusting receiver operating curves and related indices for covariates. Statistician 52, 179–192.

Food and Drug Administration Guidance for Industry, Center for Drug Evaluation and Research. (1997). In vivo bioequivalence based on population and individual bioequivalence approaches. Draft October 1997.

Food and Drug Administration Guidance for Industry, Center for Drug Evaluation and Research. (2001). Statistical approaches to establishing bioequivalence guidance for industry, Rockville, MD.

Food and Drug Administration Guidance for Industry, Center for Drug Evaluation and Research. (2003). Bioavailability and bioequivalence for orally administered drug products-general considerations, guidance for industry, Rockville, MD.

Franklin, L. A., and Wasserman, G. S. (1991). Bootstrap confidence interval estimates of C_{pk}: An introduction. Commun. Stat. Simul. Comput. 20, 231–242.

Franklin, L. A., and Wasserman, G. S. (1992). Bootstrap lower confidence limits for capability indices. J. Qual. Technol. 24, 196–210.

Franklin, L. A., and Wasserman, G. S. (1994). Bootstrap lower confidence limit estimates of Cjkp. (the new flexible process capability index). Pakistan J. Stat. A 10, 33–45.

Freedman, D. A., and Peters, S. C. (1984). Bootstrapping a regression equation: Some empirical results. J. Am. Stat. Assoc. 79, 97–106.

Gnanadesikan, R. (1977), Methods for Statistical Data Analysis of Multivariate Observations. Wiley, New York.

Gong, G. (1982). Some ideas to using the bootstrap in assessing model variability in regression. Proc. Comput. Sci. Stat. 24, 169–173.

Gong, G. (1986). Cross-validation, the jackknife, and bootstrap. Excess error in forward logistic regression. J. Am. Stat. Assoc. 81, 108–113.

Gunter, B. H. (1989a). The use and abuse of C_{pk}. Qual. Prog. 22 (3), 108–109.

Gunter, B. H. (1989b). The use and abuse of C_{pk}. Qual. Prog. 22 (5), 79–80.

Gunter, L., Zhu, J., and Murphy, S. A. (2011). Variable selection for qualitative interactions. Stat. Methodol. 8, 42–55.

Hall, P. (1985). Resampling a coverage pattern. Stoc. Proc. 20, 231–246.

Hall, P. (1988a). Theoretical comparison of bootstrap confidence intervals. Ann. Stat. 16, 927–985.

Hall, P. (1988b). Introduction to the Theory of Coverage Processes. Wiley, New York.

Hasselblad, V. (1966). Estimation of parameters for a finite mixture of normal distributions. Technometrics 8, 431–444.

Hasselblad, V. (1969). Estimation of finite mixtures of distributions from the exponential family. J. Am. Stat. Assoc. 64, 1459–1471.

Hauck, W. W., and Anderson, S. (1991). Individual bioequivalence: What matters to the patient. Stat. Med. 10, 959–960.

Hauschke, D., Steinijans, V., and Pigeot, I. (2007). Bioequivalence Studies in Drug Development Methods and Applications. Wiley, Chichester.

Heavlin, W. D. (1988). Statistical properties of capability indices. Technical Report No. 320, Advanced Micro Devices, Inc., Sunnyvale, CA.

Hsu, J. C. (1996). Multiple Comparisons: Theory and Methods. Chapman & Hall, London.

Hyde, J. (1980). Testing survival with incomplete observations. In Biostatistics Casebook. (R. G. Miller Jr., B. Efron, B. Brown, and L. Moses, eds.). Wiley, New York.

Jaki, T., Wolfsegger, M. J., and Lawo, J.-P. (2010). Establishing bioequivalence in complete and incomplete data designs using AUCs. J. Biopharm. Stat. 20, 803–820.

Kane, V. E. (1986). Process capability indices. J. Qual. Technol. 24, 41–52.

Kaplan, E. L., and Meier, P. (1958). Nonparametric estimation from incomplete samples. J. Am. Stat. Assoc. 53, 457–481.

Kocherlakota, S., Kocherlakota, K., and Kirmani, S. N. U. A. (1992). Process capability indices under non-normality. Int. J. Math. Statist. 1, 175–209.

Kotz, S., and Johnson, N. L. (1993). Process Capability Indices. Chapman & Hall, London.

Kotz, S., and Lovelace, C. R. (1998). Process Capability Indices in Theory and Practice. Arnold, London.

Kunsch, H. (1989). The jackknife and the bootstrap for general stationary observations. Ann. Stat. 17, 1217–1241.

Lahiri, S. N. (1996). On Edgeworth expansions and the moving block bootstrap for studentized M-estimators in multiple linear regression models. J. Mult. Anal. 56, 42–59.

Lahiri, S. N. (2003). Resampling Methods for Dependent Data. Springer-Verlag, New York.

Lahiri, S. N. (2006). Bootstrap methods: A review. In Frontiers in Statistics (J. Fan, and H. K. Koul, eds.). Imperial College Press, London.

Lavori, P. W., Dawson, R., and Shera, D. (1995). A multiple imputation strategy for clinical trials with truncation of patient data. Stat. Med. 14, 1913–1925.

Leadbetter, M. R., Lindgren, G., and Rootzen, H. (1983). Extremes and Related Processes of Random Sequences and Processes. Springer-Verlag, New York.

Little, R. J. A., and Rubin, D. B. (1987). Statistical Analysis with Missing Data. Wiley, New York.

Mardia, K. V., Kent, J. T., and Bibby, J. M. (1979). Mutivariate Analysis. Academic Press, London.

Martin, M., and Roberts, S. (2006). An evaluation of bootstrap methods for outlier detection in least squares regression. J. Appl. Statist. 33, 703–720.

Matheron, G. (1975). Random Sets and Integral Geometry. Wiley, New York.

McLachlan, G. J. (1987). On bootstrapping the likelihood ratio test statistic for the number of components in a mixture. Appl. Stat. 36, 318–324.

McLachlan, G. J., and Basford, K. E. (1988). Mixture Models, Inference, and Applications to Clustering. Marcel Dekker, New York.

McLachlan, G. J., and Krishnan, T. (1997). The EM Algorithm and Its Extensions. Wiley, New York.

Meeker, W. Q., and Escobar, L. A. (1998). Statistical Methods for Reliability Data. Wiley, New York.

Miller, R. G. Jr. (1981). Simultaneous Statistical Inference, 2nd Ed. Springer-Verlag, New York.

Miller, A. J. (1990). Subset Selection in Regression. Chapman & Hall, London.

Morris, M. D., and Ebey, S. F. (1984). An interesting property of the sample mean under a first-order autoregressive model. Am. Stat. 38, 127–129.

Munk, A., and Czado, C. (1999). A completely nonparametric approach to population bioequivalence in crossover trials. Preprint No. 261, Preprint Series of the Faculty of Mathematics, Ruhr-Universitat, Bochum.

Murthy, V. K. (1974). The General Point Process: Applications to Structural Fatigue, Bioscience and Medical Research. Addison-Wesley, Boston, MA.

Noreen, E. (1989). Computer-Intensive Methods for Testing Hypotheses. Wiley, New York.

Opdyke, J. D. (2010). Much faster bootstraps using SAS? InterStat. October 2010 (has code for five SAS routines used in the comparison but does not contain the two hashing algorithms).

Patterson, S., and Jones, B. (2006). Bioequivalence and Statistics in Pharmacology. Chapman & Hall/CRC, Boca Raton, FL.

Pigeot, I. (2001). The jackknife and bootstrap in biomedical research—Common principles and possible pitfalls. Drug Inf. J. 35, 1431–1443.

Price, B., and Price, K. (1992). Sampling variability of capability indices. Wayne State University Technical Report, Detroit.

Reid, N. (1981). Estimating the median survival time. Biometrika 68, 601–608.

Resnick, S. I. (1987). Extreme Values, Regular Variation, and Point Processes. Springer-Verlag, New York.

Ripley, B. D. (1981). Spatial Statistics. Wiley, New York.

Ripley, B. D. (1988). Statistical Inference for Spatial Processes. Cambridge University Press, Cambridge.

Rissanen, J. (1989). Stochastic Complexity in Statistical Inquiry. World Scientific Publishing Company, Singapore.

Rubin, D. B. (1981). The Bayesian bootstrap. Ann. Stat. 9, 130–134.

Rubin, D. B. (1987). Multiple Imputation for Nonresponse in Surveys. Wiley, New York.

Ryan, T. (1989). Statistical Methods for Quality Improvement. Wiley, New York.

Sauerbrei, W., and Schumacher, M. (1992). A bootstrap resampling procedure for model building: Application to the Cox regression model. Stat. Med. 11, 2093–2109.

Schall, R., and Luus, H. G. (1993). On population and individual bioequivalence. Stat. Med. 12, 1109–1124.

Schall, R., Endrenyi, L., and Ring, A. (2010). Residuals and outliers in replicate design crossover studies. J. Biopharm. Stat. 20, 835–849.

Schisterman, E., Faraggi, D., and Reiser, B. (2004). Adjusting the generalized ROC curve for covariates. Stat. Med. 23, 3319–3331.

Shao, J., Kubler, J., and Pigeot, I. (2000). Consistency of the bootstrap procedure in individual bioequivalence. Biometrika 87, 573–585.

Singh, K. (1981). On the asymptotic accuracy of Efron's bootstrap. Ann. Stat. 9, 1187–1195.

Singh, K., and Xie, M. (2003). Bootlier-plot—Bootstrap based outlier detection plot. Sankhya 65, 532–559.

Solow, A. R. (1985). Bootstrapping correlated data. J. Int. Assoc. Math. Geol. 17, 769–775.

Srivastava, M. S., and Lee, G. C. (1984). On the distribution of the correlation coefficient when sampling from a mixture of two bivariate normal distributions. Can. J. Stat. 2, 119–133.

Stein, M. L. (1987). Large sample properties of simulations using Latin hypercube sampling. Technometrics 29, 143–151.

Stein, M. L. (1989). Asymptotic distributions of minimum norm quadratic estimators of the covariance function of a Gaussian random field. Ann. Stat. 17, 980–1000.

Stein, M. L. (1999). Interpolation of Spatial Data: Some Theory for Kriging. Springer-Verlag, New York.

Todem, D., Fine, J., and Peng, L. (2010). A global sensitivity test for evaluating statistical hypotheses with nonidentifiable models. Biometrics 66, 558–566.

Turnbull, B. W., and Mitchell, T. J. (1978). Exploratory analysis of disease prevalence data from sacrifice experiments. Biometrics 34, 555–570.

Upton, G. J. G., and Fingleton, B. (1985). Spatial Data Analysis by Example, Vol. 1, Point Patterns and Quantitative Data. Wiley, Chichester.

Upton, G. J. G., and Fingleton, B. (1989). Spatial Data Analysis by Example, Vol. 2, Categorical and Directional Data. Wiley, Chichester.

Wasserman, G. S., Mohsen, H. A., and Franklin, L. A. (1991). A program to calculate bootstrap confidence intervals for process capability index, C_{pk}. Commun. Stat. Simul. Comput. 20, 497–510.

Westfall, P. (1985). Simultaneous small-sample multivariate Bernoulli confidence intervals. Biometrics 41, 1001–1013.

Westfall, P., and Young, S. S. (1993). Resampling-Based Multiple Testing: Examples and Methods for P-Value Adjustment. Wiley, New York.

Wolfe, J. H. (1967). NORMIX: Computational methods for estimating the parameters of multivariate normal mixtures of distributions. Research Memo. SRM 68-2, U.S. Naval Personnel Research Activity, San Diego, CA.

Wolfe, J. H. (1970). Pattern clustering by multivariate mixture analysis. Multivar. Behav. Res. 5, 329–350.

Yao, F., Craiu, R. V., and Reiser, B. (2010). Nonparametric covariate adjustment for receiver operating characteristic curves. Can. J. Stat. 38, 27–46.

8

WHEN THE BOOTSTRAP IS INCONSISTENT AND HOW TO REMEDY IT

For a very wide variety of problems, there are natural ways to bootstrap. Incredibly, most often the methods can be verified by simulation or by asymptotic theory or both. However, there are times when the bootstrap approach that is proposed fails to be consistent. In these cases, it is not obvious why the bootstrap fails and the failure is difficult to diagnose. Experience has shown that often a modification to the bootstrap can improve it.

An example is the 632 estimator of error rate in discriminant analysis. It is a case where the bootstrap works, but a less obvious modification works better. Then, there are situations where the "naïve" bootstrap is inconsistent. Examples of this include the estimation of extreme values, the estimation of a mean when the variance is infinite, and a bootstrap approach to individual bioequivalence. In each of the cases above, there is a modification that makes the bootstrap consistent. A possibly surprising result is that the m-out-of-n bootstrap is a very simple general modification that works as a remedy in several different cases. Although problems are hard to diagnose the jackknife-after-bootstrap is one technique that can help, and we will discuss that in Section 8.7.

An Introduction to Bootstrap Methods with Applications to R, First Edition. Michael R. Chernick, Robert A. LaBudde.
© 2011 John Wiley & Sons, Inc. Published 2011 by John Wiley & Sons, Inc.

8.1 TOO SMALL OF A SAMPLE SIZE

For the nonparametric bootstrap, the bootstrap samples are drawn from a discrete set, namely the original n values. Some exact and approximate results about the bootstrap distribution can be found using results from classical occupancy theory or multinomial distribution theory (see Chernick and Murthy, 1985 and Hall, 1992, appendix I). Even though the theory of the bootstrap applies to large samples, there are examples where it has been shown through simulation to work well in small samples.

In Chapter 2, we saw examples where bootstrap bias correction worked better than cross-validation in estimating the error rate of linear discriminant rules. That was one example where the bootstrap worked well in small samples. When the sample size is very small, we have good reason not to trust the bootstrap or almost any statistical procedure because the sample may not be representative of the population, and the bootstrap samples tend to repeat observations and so they tend to differ even more from the population sampling distribution than the original samples do.

The bootstrap distribution is defined to be the distribution of all possible samples of size n drawn with replacement from the original sample of size n. The number of possible samples or atoms is shown in Hall (1992, appendix I) to be

$$C_n^{2n-1} = (2n-1)![n!(n-1)!].$$

Even for n as small as 20, this is a very large number, and consequently, when generating the resamples by Monte Carlo, the probability of repeating a particular resample is small. For $n = 10$, the number of possible atoms is 92,378. Note that we are talking about distinct samples. These are distinct if no two observations in the original sample are repeat values. When sampling from a continuous distribution, that should be the case. It is often mentioned that there are only $n!$ permutations, but there are n^n bootstrap samples. However, the n^n bootstrap samples are not unique since a permutation of a bootstrap sample still contains the same set of observations and so is the same bootstrap sample as order does not matter for independent and identically distributed (i.i.d.) or even exchangeable samples. Hall's formula considers the repeats and thus provides the distinct samples for comparison. It is easy to see that $3^{n-1} \geq C_n^{2n-1} \geq 2^{n-1}$ for $n > 2$. So eventually, the number of permutations grows faster than the number of bootstrap atoms.

The conclusion is that for very small sample sizes like $n < 10$, the bootstrap may not be reliable, but as Hall showed for $n = 20$ using $B = 2000$ for the number of bootstrap replications, the probability is more than 0.95 that none of the bootstrap samples will be replicated. So n as small as 20 may still be adequate for bootstrapping.

8.2 DISTRIBUTIONS WITH INFINITE SECOND MOMENTS

8.2.1 Introduction

Singh (1981) and Bickel and Freedman (1981) independently using different methods proved the following: If $X_1, X_2, \ldots, X_n$ are independent and identically distributed random variables with finite second moments and satisfying the conditions for the central limit theorem and $Y_1, Y_2, \ldots, Y_n$ is a simple random sample chosen by sampling with replacement from the empirical distribution for the X's, then, letting

$$H_n(x, \varpi) = P\{(\widehat{Y}_n - \widehat{X}_n)/S_n \le x | X_1, X_2, \ldots, X_n\},$$

where

$$\widehat{Y}_n = n^{-1} \sum_{i=1}^{n} Y_i,$$

$$\widehat{X}_n = n^{-1} \sum_{i=1}^{n} X_i,$$

$$S_n^2 = n^{-1} \sum_{i=1}^{n} (X_i - \widehat{X}_n)^2,$$

and $\Phi(x)$ is the cumulative standard normal distribution, we have

$$\sup |H_n(x, \varpi) - \Phi(x)| \to 0 \quad -\infty < x < \infty \text{ with probability } 1.$$

Here, ϖ denotes the random outcome that leads to the values $X_1, X_2, \ldots, X_n$, and $H_n(x, \varpi)$ is a random conditional probability distribution. For fixed values of $X_1, X_2, \ldots, X_n$ and a fixed ϖ, $H_n(x, \varpi)$ is a cumulative probability function. By the central limit theorem,

$$G_n(x) = P\{[(\widehat{X}_n - \mu)/\sigma] \le x\} \to \Phi(x).$$

The bootstrap principle replaces μ and σ with $\widehat{X}_n$ and S_n, respectively, and $\widehat{X}_n$ with $\widehat{Y}_n$ in the equation above. Bickel and Freedman (1981) and Singh (1981) proved the above result for the bootstrap and hence established the consistency of the bootstrap mean.

Athreya (1987) raised the question as to whether the bootstrap remains consistent when the variance is infinite, but appropriately normalized, the sample mean converges to a stable distribution. The stable laws are described in detail in Feller (1971). Briefly, if the X_i's have a cumulative distribution F satisfying

$$1 - F(x) \sim x^{-\alpha} L(x)$$

and $f(-x) \sim cx^{-\alpha} L(x)$ as $x \to \infty$ where $L(x)$ is a slowly varying function and f is the density function (i.e., the derivative of F) and c is a non-negative constant and $0 < \alpha \le 2$. F is called stable because under these conditions and an appropriate normalization, the sample mean converges to a stable law with the distribution a function of α.

8.2.2 Example of Inconsistency

Athreya (1987) proved the inconsistency of the sample mean in the infinite variance case when a stable law applies and that is for all $0 < \alpha < 2$. In theorem 1, he only proved it for $1 < \alpha < 2$. He referred to an unpublished report for the cases when $0 < \alpha \leq 1$. The essence of the theorem is that appropriately normalized sample mean converges to a stable distribution. But the bootstrap mean when given the same normalization with the usual "bootstrap principle" substitutions converges to a random probability distribution and hence not the stable distribution. This then implies that the bootstrap is not consistent in this case.

Mathematically, this is described as follows:

$$H_n(x, \varpi) = P[T_n \leq x | X_1, X_2, \ldots, X_n],$$

where

$$T_n = nX_{(n)}^{-1}(\widehat{Y}_n - \widehat{X}_n)$$

and

$$X_{(n)} = \max(X_1, X_2, \ldots, X_n),$$

and let $G(x)$ be the limiting stable distribution for the mean. For the bootstrap to be consistent, we would need

$$\sup |H_n(x, \varpi) - G(x)| \to 0$$

for all ϖ except for a set of ϖ with measure 0:

$$-\infty < x < \infty.$$

But since Athreya showed that $H_n(x, \varpi)$ converges to a random probability measure, it cannot converge to G. A similar result was shown for the maximum by Bickel and Freedman (1981) and Knight (1989). Angus generalized the result to show it holds for minima as well as maxima. This is covered in the next section.

8.2.3 Remedies

In his paper, Athreya pointed out two remedies to attain consistency. One would be to replace the sample mean with a trimmed mean. By appropriate trimming, we get consistency for the bootstrap, but this seems to sidestep the problem because the trimmed mean does not necessarily converge to the population mean unless the distribution is symmetric, trimming is equal on both sides, and the proportion trimmed m goes to infinity more slowly than n. The second approach does not require any assumptions like those stated above for the trimmed mean. This second approach is the m-out-of-n bootstrap that we have encountered previously. Athreya proved consistency when m and n both go to ∞ at rates so the $m/n \to 0$.

8.3   ESTIMATING EXTREME VALUES

8.3.1   Introduction

Suppose we have a sample of size n from a probability distribution F, which is in the domain of attraction of one of the three extreme value types. Conditions for this are given in the theorem of Gnedenko. See Galambos (1978) for the details. All that is important here is that you know that the tail properties of F determine the norming constants and the limiting type. The theorem can be applied to the minimum or maximum of a sequence.

So, just as with the central limit theorem and the stable laws, the maximum appropriately normalized converges to a known distribution G as the sample size $n \to \infty$ for an independent identically distributed sequence of random variables. Just as in the other cases, we wish to apply the substitutions due to the bootstrap principle and see if the normalized bootstrap distribution for the maximum also converges to that same distribution G.

8.3.2   Example of Inconsistency

Mathematically, for the original data,

$$P\big[(X_{(n)} - a_n)/b_n \le t\big] \to G(t).$$

For the bootstrap samples, the substitution causes the norming constants to change to say a_n^* and b_n^*. Then, for consistency, we must have

$$P\big[(Y_{(n)} - a_n^*)/b_n^* \le t\big] \to G(t).$$

But Angus showed that

$$P\big[(Y_{(n)} - a_n^*)/b_n^* \le t\big] \to Z(t),$$

where

$$P[Z \le t] = (e-1)\sum_{r=1}^{\infty} e^{-r} P[Z_r \le t]$$

and Z_r is the rth order statistic from the original data.

8.3.3   Remedies

The m-out-of-n bootstrap provides a remedy here in a similar way as with the mean. A proof of this can be found in Fukuchi (1994). See Bickel et al. (1997) and Politis et al. (1999) for a greater appreciation of the m-out-of-n bootstrap.

Zelterman (1993) provided a different approach to remedy the inconsistency. He used a semiparametric method and was the first to find a remedy for this.

8.4 SURVEY SAMPLING

8.4.1 Introduction

In survey sampling, the target population is always finite. Assume the population size is N and the sample size is n and that n/N is not very small. Under these conditions, the variance of estimates such as sample averages is smaller than what it would be based on the standard theory of an infinite population. The finite population size also induces a correlation between observations that is negative but may be small.

Recall for independent identically distributed random variables drawn from an infinite population with a population mean μ and variance σ^2, the sample mean $\hat{X}$ has mean μ and variance σ^2/n where n is the sample size. Now, suppose we take a simple random sample of size n from a finite population of size N. Let us also assume that the finite population has mean μ and variance σ^2. Then what is the mean and variance for the sample mean?

We will not go through the computations here. The work of Cochran (1977, pp. 23–24) is a very good source for the derivation of the result for the sample mean. The mean of the sample average is also μ, but the variance is $(\sigma^2/n)(N - n)/N$ or $(1 - f)$ (σ^2/n) where $f = n/N$ is called the finite correction factor. Since $0 < f \leq 1$, the variance is reduced by a factor of $(1 - f)$.

8.4.2 Example of Inconsistency

Now, if we apply the ordinary bootstrap to the sample doing simple random sampling with replacement from the original sample, the bootstrap will mimic independence and the variance estimate for the mean will be larger than the correct variance $(1 - f)(\sigma^2/n)$. So in this sense, the bootstrap fails because it does not get the distribution of the sample mean right. Asymptotic theory for finite populations require n and N to both tend to infinity but with n at a slower rate than N or proportional to N with proportionality, possibly a constant $c < 1$. It is clear in the latter case that the bootstrap variance is wrong.

8.4.3 Remedies

Since f is known, we can apply the factor $(1 - f)$ to the bootstrap estimate to reduce the variance by the correct amount. The m-out-of-n bootstrap will work if we take m to be appropriately namely, $m/n = n/N$ or $m = n^2/N$ or even m approaching n^2/N as m, n, and N all tend to infinity. Note that normally, if we take the m-out-of-n at a rate so that m is $o(n)$ as is usually done, we will not get the correct variance, but if we take m as $O(n)$ such that $m/n \to f$, we will have the consistency. The approach can be extended to stratified random sampling from a finite population. If f is small, say, <0.01, then the effect on the variance is also small. However, Davison and Hinkley (1997) pointed out that in many practical situations, f can be in the range 0.1–0.5, which makes the finite correction very necessary.

8.5 *M*-DEPENDENT SEQUENCES

8.5.1 Introduction

An infinite sequence of random variables $\{X_i\}$ for $i = 1, 2, 3, 4, \ldots, n, \ldots \infty$ is called "m-dependent" if, for every i, X_i and X_{i+m-1} are dependent but X_i and X_{i+m} are independent. One example of this is the mth order moving average time series model. It is defined by the following equation:

$$X_i - \mu = \sum_{i=1}^{m} \alpha_i \varepsilon_{j-i} \quad \text{for} \quad j = 1, 2, \ldots, m, \ldots, \infty.$$

The process is stationary and μ is the mean of the process. The ε_j's are called innovations and are independent identically distributed, and the α_i's are the moving average parameters.

This stationary time series is m-dependent because the dependence between neighboring observations is determined by the ε's that they have in common. When they are separated by more than m points in time, they share none of the ε's. So they are then independent. If we have an m-dependent process and use the naïve bootstrap as though the observations were i.i.d., we will see that the bootstrap estimate is inconsistent.

8.5.2 Example of Inconsistency When Independence Is Assumed

Singh (1981) along with Bickel and Freedman (1981) were the first to prove that the bootstrap estimate of the mean is consistent in the i.i.d. case when the population distribution satisfies conditions for the central limit theorem. There is also a central limit theorem for m-dependent processes. See Lahiri (2003, theorem A.7 in appendix A). However, the variance for the sample mean of n consecutive observations from an m-dependent process is not σ^2/n as it would be in the i.i.d. case but rather is as follows assuming the time series is m-dependent with each X_i having a finite mean μ and finite variance σ^2.

Let $\sigma_{n,m}^2$ = variance of the mean for n consecutive observations from an m-dependent time series as defined above.

Then,

$$\sigma_{n,m}^2 = \left[n\,var(X_1) + 2\sum_{i=1}^{m-1} cov(X_1, X_{1+i}) \right] \Big/ n^2 = \sigma^2/n + 2\sum_{i=1}^{m-1} cov(X_1, X_{1+i}) \Big/ n^2.$$

If all the α's are positive, then $\sigma_{n,m}^2 > \sigma^2/n$, and if all the α's are negative, $\sigma_{n,m}^2 < \sigma^2/n$.

Regardless, in general, $\sigma_{n,m}^2 \neq \sigma^2/n$. So again, the bootstrap is getting the variance wrong and hence will be inconsistent.

8.5.3 Remedy

Since the only problem preventing consistency is that the normalization is wrong, a remedy would be to simply correct the normalization. This is achieved by using a sample estimate of $\sigma_{n,m}^2$.

8.6 UNSTABLE AUTOREGRESSIVE PROCESSES

8.6.1 Introduction

An unstable autoregressive process of order p is given by the equation

$$X_j = \sum_{i=1}^{p} \rho_i X_{j-I} + \varepsilon_i, \quad \text{for } j \in \mathbb{Z},$$

where the $\{\varepsilon_i\}$ for $i = 1, 2, 3, 4, \ldots, n, \ldots, \infty$ is the innovation process and at least one of the roots of the characteristic polynomial is on the unit circle with the remaining roots inside the until circle. In the next subsection, we will show that the stationary bootstrap previously defined is inconsistent in the unstable case.

8.6.2 Example of Inconsistency

For the unstable autoregressive process, the autoregressive bootstrap (ARB) defined for stationary autoregressive processes is inconsistent when applied to unstable autoregressive process. To understand the problem, consider the simplest case when $\rho = 1$ we have

$$X_j = \rho_1 X_{j-1} + \varepsilon_j, \quad \text{for each } j \in \mathbb{Z}.$$

Now, to be unstable, the root to the characteristic polynomial $\Psi(z) = z - \rho_1$ on the unit circle. Since the polynomial has degree 1, the root must be a real number, so $|\rho_1| = 1$ and hence ρ_1 can only be -1 or $+1$. Lahiri (2003, p. 209) pointed out that the least squares estimate of ρ_1 is consistent in this case but with a different rate of convergence and a limiting distribution that is different from both the stationary case and the explosive case.

We will not go through Datta's proof of the inconsistency for the general pth order unstable AR process when the resample size $m = n$, where n is the original length of the series. The interested reader can see Datta (1996) for the proof. Similar to other examples previously mentioned, the normalized estimate converges to a random measure rather than the fixed limiting distribution that applies in the stationary case.

8.6.3 Remedies

Just as in other cases described in this chapter, an m-out-of-n bootstrap provides an easy fix. Datta (1996) and Heimann and Kreiss (1996) independently showed that the

ARB with a resample size m approaches infinity at a slower rate than n is consistent. There are slight differences in the theorems as Datta assumed the existence of $2 + \delta$ moments for the innovations, while Heimann and Kreiss (1996) only assumed second moments for the innovations. The more stringent assumption by Datta enables him to prove almost sure convergence, whereas Heimann and Kreiss can only show convergence in probability, which is a weaker form of convergence.

Another approach due to Datta and Sriram (1997) use a shrinkage estimate for the parameters rather than the least squares estimate and bootstrapping this estimate. They were able to show that their modification of ARB is consistent.

For the m-out-of-n bootstrap, there is always an issue as to how to choose m in practice for any fixed n. In a special case, Datta and McCormick (1995) chose m using a version of the jackknife after the bootstrap method. Sakov and Bickel (1999) also in more general terms addressed the choice of m when m-out-of-n bootstrap is used.

8.7 LONG-RANGE DEPENDENCE

8.7.1 Introduction

The stationary *ARMA* processes described previously are weak or short-range dependent because their correlation function after a certain number of lags decays to 0 exponentially. Other stationary processes are called long-range dependent because their autocorrelation function decays to 0 much more slowly than the others. This is well defined by the mathematic condition that sum of the autocorrelations over all lags is a divergent series. This is one mathematical characterization, but there are other characterizations that are used in the literature. Examples can be found in Beran (1994) and Hall (1997).

8.7.2 Example of Inconsistency

Theorem 10.2 in Lahiri (2003, p. 244) shows that in the case of long-range dependence, the moving block bootstrap (MBB) is inconsistent for the normalized sample mean. The problem is that the rate of convergence is slower for long-range dependent processes, and so the norming constant used for short range causes the distribution to degenerate.

8.7.3 A Remedy

The appropriate remedy in this case is to modify the norming constants to get a nondegenerate limit. See Lahiri (2003, theorem 10.3) to see how to revise the normalization.

8.8 BOOTSTRAP DIAGNOSTICS

In parametric problems such as linear regression, there are many assumptions that are required to make the methods work well (e.g., for the normal linear models, [1] homogeneity of variance, [2] normally distributed residuals, and [3] no correlation or trend among the residuals). These assumptions can be checked using various statistical tests or through the use of graphical diagnostic tools. Similarly, diagnostics can be applied in the case of parametric bootstrap.

However, in the nonparametric setting, the assumptions are minimal, and it is not clear how to tell when a proposed bootstrap method will work properly. Many statisticians thought that diagnostics for bootstrap failure would not be possible.

Efron (1992) introduced the idea of a jackknife-after-bootstrap approach to help determine bootstrap failure. The idea is to see how each observation from the original sample affects the bootstrap calculation if we assumed it was not part of the original sample. This addresses the following question: Once a bootstrap estimate is made, how different would the result be if a particular observation were not part of the sample?

This could be done simply by doing the bootstrap procedure with one observation left out of the sample and computing the difference from the original bootstrap. This must then be repeated for each observation. This is a highly computer-intensive method. Fortunately, this brute force approach is not necessary because we can do something computationally equivalent but not as burdensome.

As we saw in Chapter 2, if the sample size is large and we do say 50,000 bootstrap samples, we will find that on average, approximately 36.8% of the original observations will be missing from the individual bootstrap samples. Another way to look at this is that for any particular observation, approximately 36.8% of the bootstrap samples will not contain it. To be explicit, consider that the bootstrap is estimating the bias of an estimator t. Say that B is the bootstrap estimate of the bias based on resampling from the full sample. Now for each j, consider the subset N_j where the jth observation is missing.

Now, consider the bootstrap estimate of bias that is obtained by restricting the calculation to just the N_j samples with the jth observation missing. Denoted by B_{-j}, the bias is computed on the subset N_j only. The value $n(B_{-j} - B)$ is the jackknife estimate of the effect of observation X_j on the bootstrap estimate of bias. Each of these estimates is like a pseudovalue. This can be done for each j. Each value is akin to the empirical influence function for the observation X_j on the estimator. The name jackknife after bootstrap came about because it describes the process. A jackknife estimate of the effect of the observations on the bootstrap estimate is evaluated after computing the bootstrap estimate. Plots have been devised to show the effect of the bootstrap. See Davison and Hinkley (1997, pp. 114–116).

8.9 HISTORICAL NOTES

Bickel and Freedman (1981) and Knight (1989) considered the example of the bootstrap distribution for the maximum of a sample from a uniform distribution on [0, θ]. Since the parameter θ is the upper end point of the distribution, the maximum $X_{(n)}$ increases to θ as $n \to \infty$; this is a case where the regularity conditions fail. They showed that the bootstrap distribution for the normalized maximum converges to a random probability measure. This is also discussed in Schervish (1995, example 5.80, p. 330).

So Bickel and Freedman (1981) provided one of the earliest examples of inconsistency for the bootstrap. Angus (1993) showed that the inconsistency applies to the maximum and minimum for any i.i.d. sequence of random variables from a distribution in the domain of attraction for one of the three extreme value types. Athreya (1987) established in general the bootstrap estimate of the sample mean is inconsistent when the variance is infinite and the distribution of the observations satisfies the conditions for a stable law. Knight (1989) provided an alternative proof of Athreya's result.

A good account of the extreme value theory in the dependent case can be found in Leadbetter et al. (1983), Galambos (1978, 1987), and Resnick (1987). Reiss (1989, pp. 220–226) presented the application of the bootstrap to estimate quantiles of the distribution of the sample. Reiss and Thomas (1997, pp. 82–84) discussed bootstrap confidence intervals.

In his monograph, Mammen (1992) attempted to show when the bootstrap can be relied on based on asymptotic theory and simulation results. A conference in Ann Arbor, MI, in 1990 was held to have the top researchers in the field explore the limitations of the bootstrap. Several of the papers from that conference were published in the book edited by LePage and Billard (1992).

Combinatorial results from the classical occupancy were used by Chernick and Murthy (1985) to establish some properties of bootstrap samples. The theoretical results can be found in Feller (1971) and Johnson and Kotz (1977). Hall (1992, appendix I) discussed the atoms of the bootstrap distribution. Also, general combinatorial results applicable to the bootstrap distribution can be found in Roberts (1984).

Regarding sampling from a finite population, the work of Cochran (1977) is one of the classic texts. Variance estimation by balanced subsampling goes back to McCarthy (1969). The early work on applying the bootstrap in the finite population context is Gross (1980). Booth et al. (1994) described the construction of studentized bootstrap confidence limits in the finite population context.

A comprehensive account of both jackknife and bootstrap approaches to survey sampling can be found in Shao and Tu (1995, chapter 6). Kovar (1985, 1987) and Kovar et al. (1988). They all performed simulations to one-stage simple random sample with replacement. Shao and Tu (1995) provided a summary of these studies and their findings. Shao and Sitter (1996) applied the bootstrap to imputation problems in the survey sampling context.

Excellent coverage of the bootstrap for stationary, unstable, and explosive autoregressive processes can be found in Lahiri (2003). Important work related to whether or not bootstrap estimated are consistent or inconsistent include Datta (1995, 1996), Datta and McCormick (1995), and Datta and Sriram (1997).

The jackknife-after-bootstrap approach to diagnostics was introduced in Efron (1992). Beran (1997) based some graphical diagnostic methods for the reliability of the bootstrap on asymptotics. Based on the regression diagnostic approach of Cook and

Weisberg (1994), Davison and Hinkley (1997, p. 119) generated linear plots for jackknife-after-bootstrap estimates.

8.10 EXERCISE

1. Surimi data: Example 3.1: Using the surimi data of Example 3.1, make a jackknife-after-bootstrap plot for the mean of the data. Hint: Use the R script:

```
require('boot')
set.seed(1) #for reproducibility
x<- c(41.28, 45.16, 34.75, 40.76, 43.61, 39.05,
41.20, 41.02, 41.33, 40.61,
  40.49, 41.77, 42.07, 44.83, 29.12, 45.59, 41.95,
45.78, 42.89, 40.42,
  49.31, 44.01, 34.87, 38.60, 39.63, 38.52, 38.52,
43.95, 49.08, 50.52,
  43.85, 40.64, 45.86, 41.25, 50.35, 45.18, 39.67,
43.89, 43.89, 42.16) #surimi data
mean(x)
bmean<- function(x,i) mean(x[i])
b<- boot(x, bmean, R=1000)
boot.ci(b)
jack.after.boot(b)
```

REFERENCES

Angus, J. E. (1993). Asymptotic theory for bootstrapping the extremes. Commun. Stat. Theory Methods 22, 15–30.

Athreya, K. B. (1987). Bootstrap of the mean in the infinite variance case. Ann. Stat. 15, 724–731.

Beran, R. J. (1994). Seven stages of bootstrap. In 25th Conference on Statistical Computing: Computational Statistics (P. Dirschdl, and R. Ostermann, eds.), pp. 143–158. Physica-Verlag, Heidelberg.

Beran, R. J. (1997). Diagnosing bootstrap success. Ann. Inst. Stat. Math. 49, 1–24.

Bickel, P. J., and Freedman, D. A. (1981). Some asymptotic theory for the bootstrap. Ann. Stat. 9, 1196–1217.

Bickel, P. J., Gotze, F., and van Zwet, W. R. (1997). Resampling fewer than n observations: Gains, losses and remedies for losses. Stat. Sin. 7, 1–32.

Booth, J. G., Butler, R. W., and Hall, P. (1994). Bootstrap methods for finite populations. J. Am. Stat. Assoc. 89, 1282–1289.

Chernick, M. R., and Murthy, V. K. (1985). Properties of bootstrap samples. Am. J. Math. Manag. Sci. 5, 161–170.

Cochran, W. (1977). Sampling Techniques, 3rd Ed. Wiley, New York.

Cook, R. D., and Weisberg, S. (1994). Transforming a response variable for linearity. Biometrika 81, 731–737.

Datta, S. (1995). Limit theory for explosive and partially explosive autoregession. Stoch. Proc. Appl. 57, 285–304.

Datta, S. (1996). On asymptotic properties of bootstrap for AR (1) processes. J. Stat. Plan. Inference 53, 361–374.

Datta, S., and McCormick, W. P. (1995). Bootstrap inference for a first-order autoregession with positive innovations. J. Am. Stat. Assoc. 90, 1289–1300.

Datta, S., and Sriram. (1997). A modified bootstrap for autoregression without stationarity. J. Stat. Plan. Inference 59, 19–30.

Davison, A. C., and Hinkley, D. V. (1997). Bootstrap Methods and Their Applications. Cambridge University Press, Cambridge.

Efron, B. (1992). Jackknife-after-bootstrap standard errors and influence functions. J. R. Stat. Soc. B 54, 83–127.

Feller, W. (1971). An Introduction to Probability Theory and Its Applications 2, 2nd Ed. Wiley, New York.

Fukuchi, J. I. (1994). Bootstrapping extremes for random variables. PhD dissertation, Iowa State University, Ames.

Galambos, J. (1978). The Asymptotic Theory of Extreme Order Statistics. Wiley, New York.

Galambos, J. (1987). The Asymptotic Theory of Extreme Order Statistics, 2nd Ed. Kreiger, Malabar.

Gross, S. (1980). Median estimation in sample surveys. Proceedings of the Section on Survey Research Methods. American Statistical Association, Alexandria, VA.

Hall, P. (1992). The Bootstrap and Edgeworth Expansion. Springer-Verlag, New York.

Hall, P. (1997). Defining and measuring long-range dependence. In Nonlinear Dynamics and Time Series, Vol. 11 (C. D. Cutler, and D. T. Kaplan, eds.), pp. 153–160. Fields Communications.

Heimann, G., and Kreiss, J.-P. (1996). Bootstrapping general first order autoregression. Stat. Prob. Lett. 30, 87–98.

Johnson, N. L., and Kotz, S. (1977). Urn Models. Wiley, New York.

Knight, K. (1989). On the bootstrap of the sample mean in the infinite variance case. Ann. Stat. 17, 1168–1175.

Kovar, J. G. (1985). Variance estimation of nonlinear statistics in stratified samples. Methodology Branch Working Paper No. 84-052E, Statistics Canada.

Kovar, J. G. (1987). Variance estimates of medians in stratified samples. Methodology Branch Working Paper No. 87-004E, Statistics Canada.

Kovar, J. G., Rao, J. N. K., and Wu, C. F. J. (1988). Bootstrap and other methods to measure errors in survey estimates. Can. J. Stat. 16 (Suppl.), 25–45.

Lahiri, S. N. (2003). Resampling Methods for Dependent Data. Springer-Verlag, New York.

Leadbetter, M. R., Lindgren, G., and Rootzen, H. (1983). Extremes and Related Properties of Random Sequences and Processes. Springer-Verlag, New York.

LePage, R., and Billard, L. (eds.). (1992). Exploring the Limits of Bootstrap. Wiley, New York.

Mammen, E. (1992). When Does the Bootstrap Work? Asymptotic Results and Simulations. Springer-Verlag, Heidelberg.

McCarthy, P. J. (1969). Pseudo-replication: Half-samples. Int. Stat. Rev. 37, 239–263.

Politis, D. N., Romano, J. P., and Wolf, M. (1999). Subsampling. Springer-Verlag, New York.

Reiss, R.-D. (1989). Approximate Distributions of Order Statistics and Applications to Nonparametric Statistics. Springer-Verlag, New York.

Reiss, R.-D., and Thomas, M. (1997). Statistical Analysis on Extreme Values with Applications to Insurance, Finance, Hydrology and Other Fields. Birkhauser Verlag, Basel.

Resnick, S. I. (1987). Extreme Values, Regular Variation, and Point Processes. Springer-Verlag, New York.

Roberts, F. S. (1984). Applied Combinatorics. Prentice-Hall, Englewood Cliffs, NJ.

Sakov, A., and Bickel, P. J. (1999). Choosing m in the m-out-of-n bootstrap. ASA Proceedings of the Section on Bayesian Statistical Science, pp. 124–128, American Statistical Association, Alexandria, VA.

Schervish, M. J. (1995). Theory of Statistics. Springer-Verlag, New York.

Shao, J., and Sitter, R. R. (1996). Bootstrap for imputed survey data. J. Am. Stat. Assoc. 91, 1278–1288.

Shao, J., and Tu, D. (1995). The Jackknife and Bootstrap. Springer-Verlag, New York.

Singh, K. (1981). On the asymptotic accuracy of Efron's bootstrap. Ann. Stat. 9, 1187–1195.

Zelterman, D. (1993). A semiparametric bootstrap technique for simulating extreme order statistic. J. Am. Stat. Assoc. 88, 477–485.

AUTHOR INDEX

An Introduction to Bootstrap Methods with Applications to R, First Edition. Michael R. Chernick, Robert A. LaBudde.
© 2011 John Wiley & Sons, Inc. Published 2011 by John Wiley & Sons, Inc.

SUBJECT INDEX

An Introduction to Bootstrap Methods with Applications to R, First Edition. Michael R. Chernick,
Robert A. LaBudde.
© 2011 John Wiley & Sons, Inc. Published 2011 by John Wiley & Sons, Inc.

Printed and bound by CPI Group (UK) Ltd, Croydon, CR0 4YY

16/04/2025

14658368-0003